ELECTROCHEMISTRY FOR ENVIRONMENTAL PROTECTION

ELECTROCHEMISTRY FOR ENVIRONMENTAL PROTECTION

Editor
Dr. DIGUMARTI BHASKARA RAO
Secretary
Academy of Communication Culture
Education Science and Service
1-22-10 Srinivasa Nagar
Guntur—522006
Andhra Pradesh
India

DISCOVERY PUBLISHING HOUSE
New Delhi.

Published by:

Tilak Wasan

DISCOVERY PUBLISHING HOUSE PVT. LTD.
4383/4B, Ansari Road, Darya Ganj
New Delhi-110 002 (India)
Phone : +91-11-23279245, 23253475, 43596065
E-mail : discoverypublishinghouse@gmail.com
sales@discoverypublishinggroup.com
web : www.discoverypublishinggroup.com

***First Edition:* 2001**

***Reprinted:* 2020**

ISBN: 978-81-7141-619-6

Electrochemistry for Environmental Protection

Printed at:
Infinity Imaging Systems
Delhi

ACKNOWLEDGEMENTS

I

am

thankful

to

Prof. Pierre Lasserre

Dr. Rosanna Santesso

UNESCO-ROSTE, Venice

Karel Stulik

Robert Kalvoda

Authors of Articles

and

UNESCO

Bhaskar R. Digumarti

PREFACE

The importance of the world's environmental problems need not be stressed. However, it should be pointed out that any approach toward coping with these problems cannot depend merely on emotional appeals to the mankind or on isolated activities in any technological, scientific or social field. The only hope lies in concerted activities of all the departments of the human society, under the guidance of sober, rational reasoning. Natural sciences occupy one of the most important places in this respect and chemistry significantly contributes both to the understanding of the environmental processes and to the technologies for environmental cleaning.

Electrochemistry deals with chemical processes involving transfer of electric charge, i.e., with oxidation and reduction. If we look at the processes occurring in nature, we can see that most of them involve redox reactions; hence, electrochemistry should substantially contribute to the understanding of life. Moreover, measurement of thermodynamic and kinetic parameters of charge-transfer reactions and of electrical quantities characterising solutions is an indispensable too for the monitoring of the composition of the environment in space and time, which is the basis for evaluation of the state of the environment and for the formulation of strategies for its mending. Finally, electrochemical processes can be utilised technologically for removal of pollutants from various environmental matrices.

There has been a long electrochemical tradition in the Czech Republic, primarily owing to the activity of the Nobel prize winner Jaroslav Heyrovsky, leading to the establishment of many centres dealing with physicochemical, analytical and technological aspects of electrochemistry. The international significance of this basic and applied research has been reflected in the establishment of the UNESCO Laboratory of Environmental Electrochemistry that was approved by the XXIIth UNESCO General Conference in 1983. At that time, the

Laboratory was part of the J. Heyrovsky Institute of Physical Chemistry and Electrochemistry, Czechoslovak Academy of Sciences. In 1990, the Laboratory became a joint research facility of the J. Heyrovsky Institute of Physical Chemistry, the Czech Academy of Sciences, and the Department of Analytical Chemistry, Charles University. These two institutions had closely cooperated before and the merger enhanced their capabilities, both in basic research and in its applications. Of course, the Laboratory has always cooperated with many other institutions and this is reflected in the fact that some chapters of this text have co-authors who are not Laboratory members.

The present publication is a selective compilation of the results obtained by UNESCO Laboratory. It is tried to create a balanced account of all the aspects of Lab's electrochemical studies that are related to theoretical and practical problems of environmental electrochemistry; therefore, the following chapters are concerned with contributions to the study of the mechanisms of environmentally important processes, their rates and equilibria, to the monitoring and determination of environmental pollutants and to the methods of degradation of toxic compounds.

This publication presents the results obtained in the UNESCO Laboratory over recent years; it cannot be exhaustive. Nevertheless, each chapter is placed within the framework of the contemporary world research and outlines the trends and relations with other fields. In our opinion, it clearly demonstrates the links among all environmental activities and underlines the need for creative approach to the solution of environmental problems.

Karel Stulik and **Robert Kalvoda**
UNESCO Laboratory of Environmental Electrochemistry,
Charles University and the Academy of Science of the Czech Republic,
Prague, Czech Republic

CONTENTS

---

* All the articles are reproduced from the Technical Report No. 25 of UNESCO-Roste, Venice with the kind permission of UNESCO Venice Office.

1
OXYGEN - THE ESSENTIAL ENVIRONMENTAL COMPONENT - AS A CLASSICAL ELECTROCHEMICAL ELEMENT

Michael Heyrovský, Stanislav Vavřička, Lubomír Šerák and Milan Fedurco

Abstract

The article brings a survey of recent results obtained at the UNESCO Laboratory in research on electroreduction of oxygen, and of their practical applications. It was shown that oxygen molecule accepts the first electron in immediate contact with the electrode in superoxo orientation and that the detachment of the resulting superoxide anion radical from the electrode surface is the slow step in oxygen reduction to hydrogen peroxide. The electrochemical generation of superoxide anion radical was utilized for determination of superoxide dismutase in human blood. On the case of photosensitized oxidation of methanol in alkaline solution it was demonstrated that electrooxidation of hydrogen peroxide can serve for study of photodynamic effect. It was found that electroreduction of oxygen to water can be efficiently catalyzed by an inorganic hydroxo complex of iron. The Clark sensor was modified and used for measurements of oxygen concentration in various biological and medical studies, also for following the effect of polluted environment upon occurrence of inborn defects.

Key Words

molecular oxygen, electroreduction, superoxide radical anion, hydrogen peroxide, hydroxyl radical, electrooxidation, disproportionation, catalysis, Clark sensors

1.1 Introduction

No reader needs to be convinced about the paramount importance of oxygen for life on earth. Its 20.9 volume % equilibrium contents in pure dry atmosphere correspond to its approximately 10^{-3} mol dm^{-3} equilibrium concen-tration in global fresh waters and in dilute solutions of electrolytes, such as brine of seas and oceans. Every kind of pollution results in a decrease of the normal oxygen levels in the environment and hence reliable methods for the control of oxygen concentration are essential for sustaining healthy conditions on the Earth.

In metabolic processes, life-giving energy is provided by chemical reactions of various substrates with oxygen in which an oxygen molecule ultimately combines either with a carbon atom to form CO_2 or with four hydrogen atoms to give two water molecules. In photosynthesis, on the other hand, oxygen is released by green plants under the action of sunlight. Measurements of oxygen uptake or production in course of these processes provide the way how to follow and characterize them in a quantitative manner for analytical or research purposes.

For deeper understanding of the role oxygen plays in the key processes of life, biologists, chemists and physicists continued studying by various methods the properties and reactivity of this ubiquitous element since its discovery by Priestley and Scheele. In 20th century, electrochemistry has been contributing considerably towards this aim, providing methods for analytical determination of oxygen as well as introducing ways to study elementary redox reactions essential in oxygen chemistry.

In polarographic research started at the Prague Charles University almost 75 years ago, the very first current-voltage curve measured with the dropping mercury electrode displayed the two-step electroreduction of molecular oxygen in 1M NaCl. Since then till the present time the electrochemical behaviour of oxygen has been figuring among the traditional research topics of the Czech polarographic school [1-12]. The research has been pursued essentially along two lines:

a) study of elementary electrode kinetics of oxygen reactions and of reactions of their products;
b) development of methods for analytical determination of oxygen and their applications for specific purposes.

In course of time it became clear that while for fundamental research the renewed mercury electrodes of high reproducibility have their special advantage, for analytical applications highly specialized solid electrodes have to be used. The present report brings a survey of our recent contributions to these two aspects of electrochemistry of oxygen.

The oxygen molecule, in its ground state a triplet $^3O_2$, can exist in its lowest excited state in two singlet forms $^1O_2$. The one with lower energy content (95 kJ mol^{-1} above ground state), denoted as $^1\Delta_g$, has its life-time in the liquid state within the order of milliseconds and hence its properties are relatively easily accessible to experimental measurements. Research on its electrochemical behaviour is currently in progress in our laboratory.

1.2 Electroreduction of oxygen to hydrogen peroxide

Molecular oxygen in course of its reduction is capable of reacting with altogether four hydrogen atoms, or, successively, with four electrons and four protons, giving water. In biological media, water is the essential component and in course of oxygen reduction it can supply the necessary protons. The elementary course of this reaction can be conveniently followed and its detailed mechanism studied under simplified conditions in which oxygen is dissolved in an aqueous solution and the reducing agent is a mercury electrode. On mercury, as on majority of other metals, the oxygen molecule accepts the four electrons from the electrode in two steps. First, in the region of more positive potentials, it is reduced by two electrons and two protons to hydrogen peroxide:

$$O_2 + 2\,e^- + 2\,H^+ \rightarrow H_2O_2 \qquad (1)$$

In this first reduction step the bond between the two oxygen atoms remains preserved. In solutions of pH > 9 the reaction proceeds reversibly which means that in the same potential region oxygen is converted electrochemically to hydrogen peroxide and hydrogen peroxide to oxygen, both reactions occurring at a high speed so that the electric currents due to reduction of oxygen or to oxidation of hydrogen peroxide are controlled by the rate of diffusion of the respective species to the electrode, all along the current-voltage curve recorded with the dropping mercury electrode or with the hanging mercury drop electrode. With pH increasing from 9 the potential of the polarographic half-wave or of the voltammetric peak pertaining to reaction (1) shift to negative values in the same way as the reversible potentiometrically measured potential, i.e. by 60 mV per pH unit.

In solutions of pH less than 9 the reaction (1) is irreversible and the half-wave and peak potentials are constant, near the potential of the saturated calomel electrode, independent of pH. By various experimental techniques of electrode kinetics it was established that under these conditions the rate-determining step in reaction (1) is the transfer of the first electron to the oxygen molecule, i.e., that the net process (1) can be interpreted as a sequence of the following elementary phases:

$$O_2 + e^- \rightarrow O_2^- \quad \text{(slow)} \qquad (2)$$

$$O_2^- + H^+ \rightarrow HO_2 \quad \text{(fast)} \qquad (3)$$

$$HO_2 + e^- \rightarrow HO_2^- \quad \text{(fast)} \qquad (4)$$

$HO_2^- + H^+ \rightarrow H_2O_2$ (fast) (5)

This conclusion rises chemists' interest, as the motion of an electron is generally much faster than the motions of molecular or ionic species; the slowness of reaction (2) would have to be then explained by slow rate of the process of accommodation of the electron on the oxygen molecule. However, such an interpretation seems not likely for a molecule with two unpaired electrons in the two highest antibonding orbitals; moreover, reaction (2) is exothermic and favoured in aqueous media by much stronger hydration of the superoxide anion radical O_2^- than of molecular oxygen O_2.

The reduction of oxygen at the potential limit in acidic solutions occurs on a metallic surface which bears a relatively high positive charge. A corresponding cathodic current can flow only if the superoxide anion radical O_2^- undergoes the steps (3), (4), (5). On the positively charged electrode there are no H^+ ions present and the orientation of water dipoles in the interface is unfavourable for splitting off a proton for a surface reaction. In order to accomplish step (3), the negative O_2^- particle must detach itself from the positive mercury surface in the field of the double layer opposing the detachment. When the double layer changes from compact to diffuse the electrostatic hindrance of the charge separation becomes less and the flow of the negative charge from the electrode increases. In the solution away from the electrode surface, the protonations (3) and (5) occur without any difficulty and the electron transfer (4) to the strongly oxidizing radical HO_2 takes place probably by tunneling. Reaction (2) can then be subdivided into 3 successive stages :

O_2 (solution) $\rightarrow$ O_2 (metal surface) fast (2a)

O_2 (metal surface) + e^- $\rightarrow$ O_2^- (metal surface) fast (2b)

O_2^- (metal surface) $\leftrightarrow$ O_2^- (solution) slow (2c)

of which the last is the slowest one.

We have observed [13] that with gradual dilution of the acidic supporting electrolyte the oxygen reduction wave and peak shift markedly to positive potentials (***Fig. 1***), which means that when the electrode double layer changes with dilution from compact to diffuse, the oxygen reduction proceeds easier and faster. The most positive limit of oxygen half-wave or peak potential was attained in 10^{-5} mol dm^{-3} solutions of strong acids, at +0.3 V *vs.* SCE. Under such conditions, the oxygen concentration in solution is about 100 times higher than that of the protons essential for reaction (1) to take place, and the reduction current is limited by the rate of migration of the H^+ ions to the electrode in the field of diffuse double layer. At more negative potentials, the protons for oxygen reduction are supplied from water in the drawn-out negative part of the wave (***Fig. 1***).

In 10^{-5}M $HClO_4$ solution we carried out exact measurements of drop-time of the dropping mercury electrode in presence and in absence of air oxygen (***Fig. 2***). The result - shorter drop-time in oxygen solutions - indicates that the oxygen reduction occurs as a surface process in direct contact with the electrode. In more concentrated electrolyte solutions the weak effect of oxygen on drop-time is overwhelmed by the adsorption of anions.

The importance for oxygen reduction of direct contact of the molecule with metallic surface can be demonstrated by an effect of surfactants. Surface active organic anions of planar structure which get adsorbed flat on positively charged mercury surface, like Na tetrahydronaphthalene-6-sulphonate or Na sulphosalicylate, shift the voltammetric oxygen reduction peak without any distortion to negative potentials, at various rates of potential scan (***Fig. 3***).

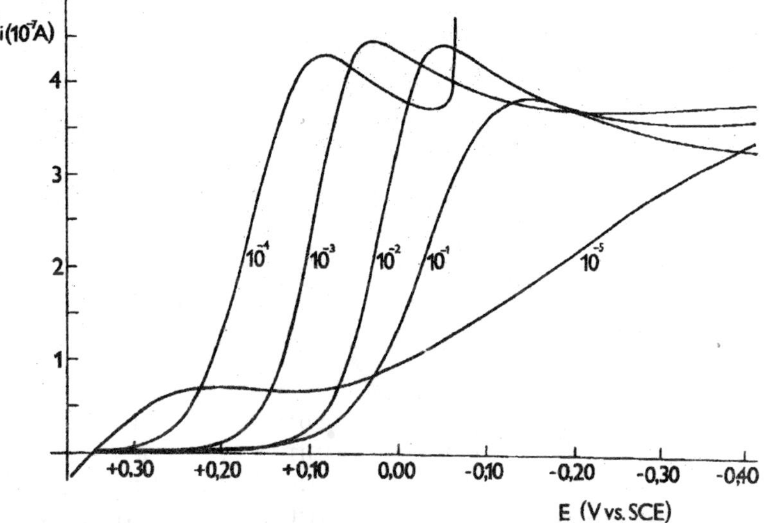

Fig. 1: **Current-potential curves of electroreduction of air oxygen to hydrogen peroxide in $HClO_4$ on a hanging mercury drop electrode**

Concentration of $HClO_4$ in mol dm^{-3} is indicated on each curve. Three-electrode system, scan rate 10 mV s^{-1}

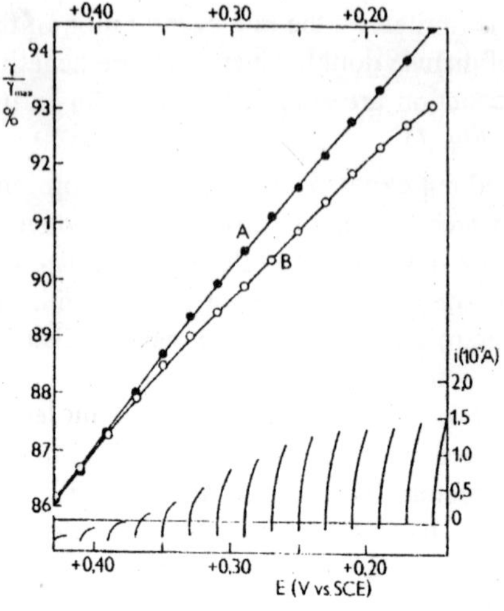

Fig. 2: **Potential dependence of the drop-time and of the current-time curves measured with the same dropping mercury electrode in the presence and in the absence of air oxygen in 10^{-5}M $HClO_4$**

Curve A - drop-time in deaerated solution; curve B - drop-time in the presence of air oxygen. Drop-time is expressed in percentage of the maximum value, t_{max} = 60.72 s, obtained at the pzc in the deaerated solution, E_{pzc} = -0.4 V.

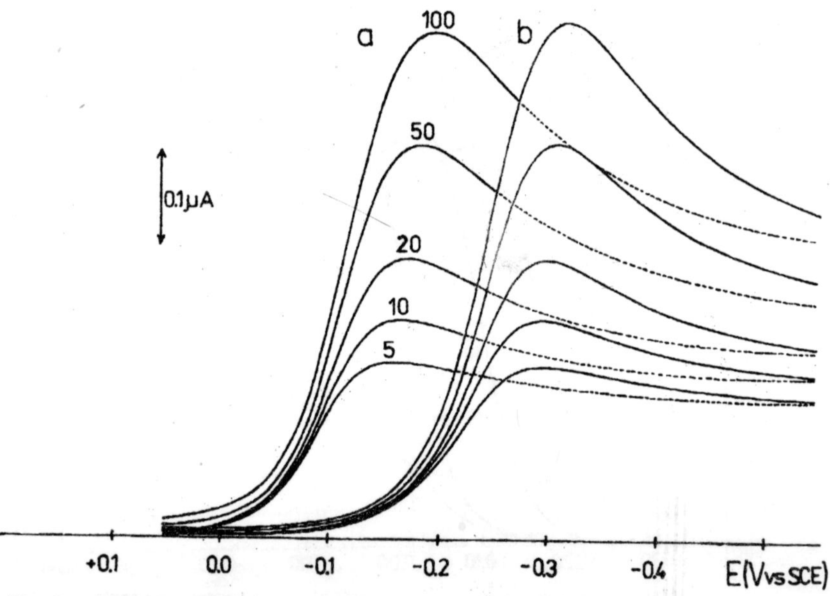

Fig. 3: **Voltammetric curves of the first reduction wave of air oxygen on a hanging mercury drop electrode recorded with rates of voltage scan of 5, 10, 20, 50, 100 mV s^{-1}**

a) in 0.1M acetate buffer of pH 4.7;

b) ditto + 2 x 10^{-3} M Na tetrahydronaphthalene-6-sulphonate

The above experimental results led us to the conclusion that for accepting electron in the energetically most advantageous way from an electrode or from other electron donors, the oxygen molecule has to come into a direct contact with the reducing partner, and they together have to provide path for electron transfer via overlapping electronic orbitals. The oxygen molecule can contact a surface either in peroxo or in superoxo orientation, depending on whether in the time of contact the bond between the two atoms is positioned in parallel or perpendicular to the surface plane. For the electron transfer (2b) to be facilitated, the favourable orientation is superoxo, with the bond perpendicular to the surface. This is analogous to the favourably oriented casual collision of molecular oxygen with an electron donor in solution on which a contact charge transfer is automatically established. Thus, electrochemistry is able to provide deeper insight into the molecular mechanism of electron transfer to oxygen which is one of the essential elementary steps of biological processes in living organisms.

1.2.1 The superoxide anion radical

Increasing pollution of the earth atmosphere, use of nitrogen-containing fertilizers, of pesticides and of medical drugs with strong oxidizing properties, high levels of radiation, stress, and the process of aging itself, represent a serious threat to the antioxidation system of a living cell, in spite of its high catalytic efficiency. For example, certain toxic chemicals might block important components of the respiratory chain (RC), directing the flow of electrons to oxygen before it reaches the last enzyme of the RC, the cytochrome c oxidase. In such cases molecular oxygen undergoes a single electron reduction yielding the superoxide anion radical $O_2^{\cdot-}$.

RC may not be the only place in eucaryotic cells where the toxic radicals such as O_2, OH and HO_2 can be formed. Some chemicals possess the ability to oxidize oxyhemoglobin (oxyHb) to methemoglobin (metHb) with a release of the O_2^- radical (see below). The superoxide, in the presence of higher levels of Fe(II), participates in the so-called Haber-Weiss reaction (HWR):

$$O_2^- + H_2O_2 \rightarrow OH + OH^- + {}^1O_2 \qquad (6)$$

One of the products of HWR, the hydroxyl radical OH, is much more reactive than O_2, and it attacks readily peptides, lipids and nucleic acids. Oxygen radicals and H_2O_2 play also an important role in leukocytes (fagocytosis and intracellular killing of microorganisms). On the other hand, superoxide and other reactive oxygen species are responsible for the initiation of lipid peroxidation, leading to hemolytic decay and early "death" of the red blood cells (RBC).

1.2.1.1 ELECTROCHEMICAL PROPERTIES OF SUPEROXIDE

The superoxide anion radical can be easily generated chemically, electrochemically or using the pulse radiolysis technique. The electrochemical method is especially convenient since the metallic electrode serves as a pure source of electrons. In addition, important parameters such as redox potentials or rate constants for O_2^- dismutation reactions, as well as information about the reaction products, can be obtained in a single electrochemical experiment. Also, the concentration of superoxide dismutase (SOD) on a sub-ppb level and its activity [14] can be measured in course of several minutes.

In aprotic solvents on Hg, Au or Pt electrodes, oxygen undergoes a one-electron reduction, yielding superoxide anion radical, which is reversibly reoxidizable to molecular oxygen. The electrode reaction is reported to be extremely fast for such a small reactant, with the standard rate constants considerably exceeding 1 cm s^{-1} in acetonitrile solutions [15]. A special property of O_2^- radical is its long-term stability in non-complexing media (at least 30 minutes in dimethylsulphoxide), as evident from ESR and electrochemical experiments [16]. Furthermore, several reports have appeared describing electrogeneration of stable O_2^- in aqueous solutions [17]. However, certain precautions have to be exercised in aqueous solutions, in order to maintain O_2^- stability during the electrochemical experiment:

a) Basicity of the solution is relatively high (pH > 9).

b) The Hg electrode has to be covered with a compact film of a surfactant impermeable to water molecules but still permeable to electrons.

c) The surfactant film should be able to block completely protonation reactions of the electrogenerated superoxide anion radical on the electrode surface.

These requirements are satisfied by molecules possessing strong surface activity in aqueous or aqueous/alcoholic solutions, e.g., α-quinoline, triphenylphosphine oxide (TPO) [15,17] or tripiperidinophosphine oxide [18].

1.2.1.2 POLAROGRAPHIC DETERMINATION OF Cu,Zn-SUPEROXIDE DISMUTASE IN HEMOLYSATES

Superoxide dismutases (SODs) are metalloenzymes present in all aerobic organisms. The SOD containing Cu and Zn is characteristic for eucaryotic organisms. It catalyses the dismutation reaction of the superoxide anion radical [19]:

$$SOD\text{-}Cu^{2+} + O_2^{\cdot -} \xrightarrow{k_1} SOD\text{-}Cu^{+} + O_2 \qquad (7)$$

$$SOD\text{-}Cu^{+} + O_2^{\cdot -} \xrightarrow{k_2} SOD\text{-}Cu^{2+} + O_2^{2-} \qquad (8)$$

$$O_2^{2-} + 2H^{+} \rightarrow H_2O_2 \qquad (9)$$

The Cu,Zn-SOD is one of the fastest enzymes known, with the rate constants k_1 and k_2 larger than 2×10^9 mol^{-1} s^{-1} (pH 7.4) [19]. In addition, because of its high stability (without a significant loss of its activity even in 8M urea or 86 % ethanol/water solution [20]), large molecular mass proteins such as hemoglobin or catalase, which would otherwise interfere with SOD determination, can be removed from RBC hemolysates using chloroform extraction method [14].

For finding the optimum concentration of TPO for the determination of SOD activity, we used logarithmic analysis of i - t curves in course of O_2 reduction in the absence and in the presence of TPO [21], and measured electrocapillary curves by means of the method of controlled convection [22]. Oxygen electroreduction on the dropping mercury electrode in the absence and in the presence of 5×10^{-4}M TPO can be seen in ***Figs. 4a,b***. The effect of Cu,Zn-SOD-containing hemolysate on the cathodic wave of oxygen with corresponding analysis of polarographic data is presented in ***Fig. 4c***. ***Figs. 4a,b*** show that the surface activity of TPO causes a complete suppression of the polarographic maximum, a 50 % decrease of the first oxygen reduction wave and its significant shift towards negative potentials. The ensuing one-electron diffusion-controlled wave then corresponds to formation of the superoxide anion radical in the vicinity of the electrode.

The presence of the superoxide is confirmed by a catalytic increment i_C of the current upon addition of SOD into the solution. Since O_2 is constantly regenerated in the $O_2^{\cdot -}$ dismutation process catalyzed by SOD (Eqs. 7, 8), the cathodic wave S increases with SOD concentration up to the two-electron diffusion-limited value M, identical with the reduction process in the absence of TPO. By definition the catalytic current increment i_C is

$$i_C = i_l - i_d \qquad (10)$$

where i_l and i_d are the values of the mean limiting current in the presence and in the absence of SOD, respectively. Since the drop-time t and the rate constant k_1 are known, the SOD content in the biological sample can be determined using the expression by Koutecky [23]:

$$X = \{7.42(i_l/i_d) - 7.42\}/\{2.25 - 1.25(i_l/i_d)\} = k_1[SOD]\, t \qquad (11)$$

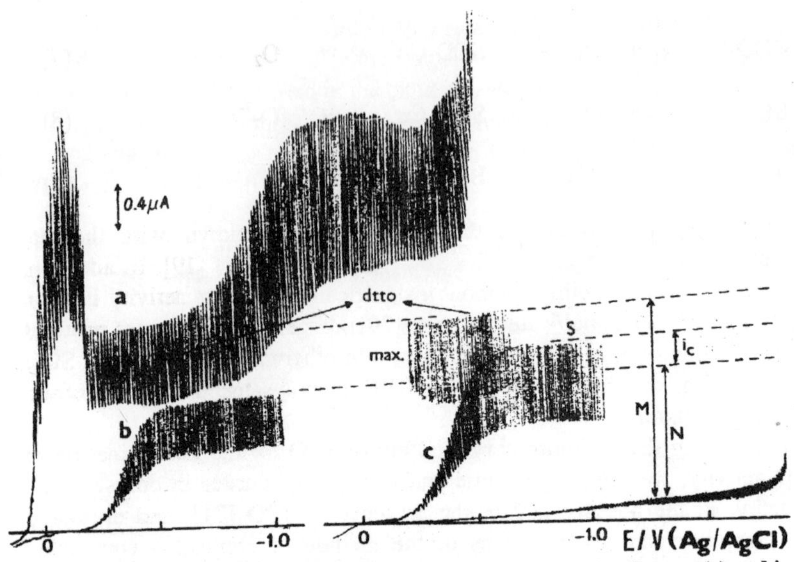

Fig. 4: **DC-polarograms of electroreduction of air oxygen in the absence (a) and in the presence (b) of 5 x 10^{-4}M TPO; (c) - effect of addition of RBC hemolysate containing SOD (discussed in the text)**
Solution - 0.025M borate buffer of pH 9.8 containing 10 volume % CH_3OH; drop-time 2 s, scan rate 5 mV s^{-1}

Tab. 1: **Dependence of i_C, i_l and Koutecky parameter X on the SOD concentration in the RBC hemolysate**

Hemolysate additions*)	i_C(μA)	i_l (μA)	X
5	0.1	0.64	1.788
10	0.18	0.72	4.240
15	0.22	0.76	6.180
20	0.26	0.80	8.971
25	0.28	0.82	10.934
50	0.30	0.84	13.491
100	0.35	0.89	25.336

i_d = 0.54 mA; scan rate 5 mVs^{-1}; t = 2 s; Hg flow rate 0.788 mg s^{-1}

*) Additions of hemolysate into 5 ml of 0.025M borate buffer containing 10 volume % CH_3OH and 5 x 10^{-4}M TPO

The results of SOD measurement are summarized in ***Table 1***. Using the rate constant $k_1 = 2.3 \times 10^9$ mol^{-1}dm^{-3} for the SOD-catalyzed superoxide dismutation [14] and *t* values in the range 0 - 11 s, one gets the slope of 0.48 which leads to 1.1×10^{-6}M SOD present in human RBC hemolyzate (after correction for its dilution in the supporting electrolyte). This value is of the same order of magnitude as 1.8×10^{-6}M SOD determined in blood samples by Rigo and Rotilio [14].

For testing the effect of nitrites two hemolysate samples were prepared. In the test sample 0.9 ml of 0.115M NaCl were pipetted to 1 ml of human RBC to which 0.1 ml of 0.155M $NaNO_2$ was added; the sample was gently stirred and kept for 15 minutes at room temperature. The reference sample contained 1 ml of 0.155M NaCl and 1 ml of human RBC. Complete hemolysis in both samples was achieved by adding 6 ml of 0.01M phosphate buffer of pH 7.4. Both samples were then centrifuged at 3000 G. To 6 ml of each sample were further added 2.4 ml of ethanol/chloroform mixture (1.5 ml EtOH + 0.9 ml chloroform) and the solutions were stirred under ice at 2 °C for 15 minutes. Finally both samples were centrifuged at 5000 G for 20 minutes and used for electrochemical experiments. The *in vitro* incubation of freshly isolated RBC with $NaNO_2$ and their subsequent hemolysis have a very interesting effect on the electrochemistry of RBC. Upon the addition of 100 µl of hemolysate into 5 ml of polarographic solution, a large increase of the cathodic limiting current is observed, caused by the SOD catalysis of O_2^- dismutation (cf. ***Figs. 5a*** and ***4b***). Besides, a small cathodic wave G appears at -0.02 V (***Fig. 5a***) which was not observed with hemolysates untreated with nitrite. The process giving rise to this wave produces a pronounced cathodic peak, denoted as G, on fast scan differential pulse voltammograms (***Fig. 5b***). A similar cathodic peak is observed on fast scan differential pulse voltammograms recorded with TPO-covered HMDEs in presence of oxidized glutathione (GSSG) (***Fig. 6***).

The effect is apparently due to the involvement of O_2 in the faradaic process of $(GS)_2Hg$ reduction on the electrode surface [24]. Similar effects were observed also in O_2 reduction in the presence of cystine on TPO-covered Hg electrodes [25]. Molecular oxygen easily penetrates to the hydrophilic sites formed upon the thiolate adsorption, and while $(RS)_2Hg$ is being reduced, O_2 accepts one electron from the reduction product - thiolate anion - and forms the superoxide. We were also able to show (SOD test) that O_2^- radical generated at about 0.0 V (***Fig. 6***) decomposes rapidly in a surface reaction with thiyl radicals, while the superoxide generated at -0.4 V (peak S) diffuses further away from the electrode surface, where it undergoes dismutation with SOD (Eqs 7,8) [26].

The cathodic wave G, for the nitrite-exposed RBC sample, might be due to higher production rates of reduced glutathione (GSH) during the incubation period of RBC with the nitrite (antioxidative protection of RBC). GSH is then further oxidized to GSSG by O_2^- and H_2O_2. It is reasonable to assume that the nitrite causes complex hemolysis of RBC, since for the same volume of added hemolysates the cathodic current in the range from -0.15 to -0.30 V is much larger in the case of the nitrite-treated sample than in the reference experiment (cf. curves H and dashed curve in ***Fig. 5b***). A complex fragmentation of -SH containing proteins cannot be excluded either. The peak denoted as K in ***Fig. 5b*** is due to a catalytic effect on oxygen reduction by hemin which released from the RBC during their incubation with NO_2^- ions.

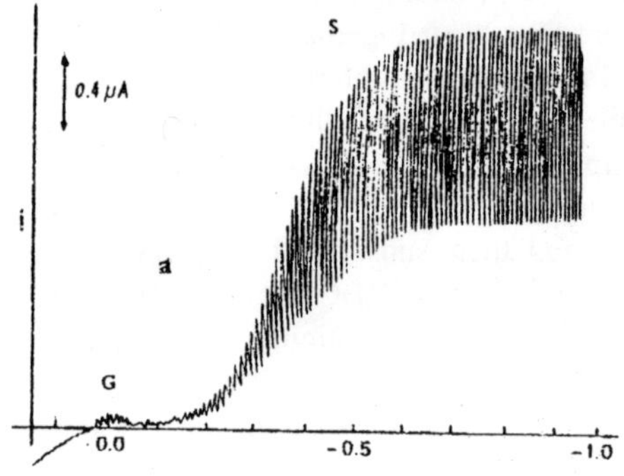

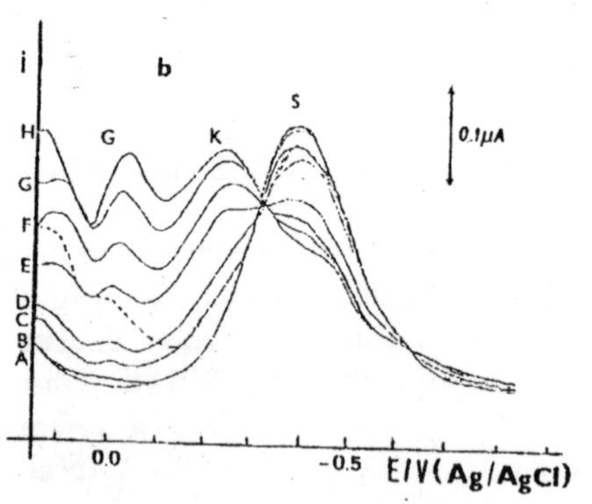

Fig. 5:

a) The same as in *Fig. 4c*, but for the hemolysate of NO_2 - treated RBC

b) Fast scan differential pulse voltammograms for oxygen reduction on TPO-covered hanging mercury drop electrode in the presenc e of increasing amount of RBC hemolysate:

A: 0; B: 5; C: 10; D: 15; E: 20; F: 25; G: 50; H: 100 μl
For comparison - dashed line corresponds to the addition of 100 μl of RBC hemolysate untreated by nitrite, while curves B - H were obtained after 15 minutes of RBC incubation with $NaNO_2$ (see text). Scan rate 20 mV s^{-1}, pulse amplitude E = 25 mV

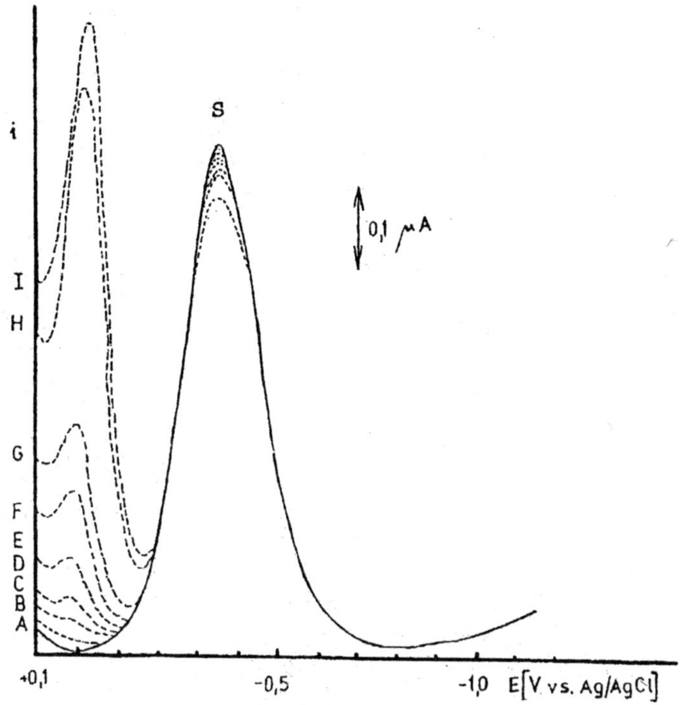

Fig. 6: **Fast scan differential pulse voltammograms for the air oxygen reduction on a hanging mercury drop electrode in the presence of increasing amount of oxidized glutathione**

Solution of 0.025M borate buffer of pH 9.8 containing 10 volume % CH_3OH, 5×10^{-4}M TPO and

A: 0; B: 1; C: 2; D: 3; E: 4; F: 5; G: 6; H: 14; I: 28 μl GSSG

Such complicated changes in the voltammetric behaviour are a clear result of the damage in RBC caused by oxygen radicals formed in the reaction of oxyHb with NO_2^- anions. The exact reaction mechanism is not completely understood [27]. However, it is highly probable that the superoxide anion radical gets further involved in the HWR (Eq. 6), and the oxygen radicals subsequently cause an irreversible damage of the RBC membranes. For the same reason, relatively low levels of nitrites in drinking water or vegetable juices are known to be deadly for infants. Their fetal oxyHb is much more likely to get oxidized to metHb by nitrite than the sterically better protected hemoglobin in adults [28].

1.2.2 Electrooxidation of hydrogen peroxide to oxygen

Hydrogen peroxide is formed as a product either of reduction of molecular oxygen or of disproportionation of superoxide anion (Eqs 7, 8). As mentioned in part 1.2, the electroreduction of oxygen to hydrogen peroxide proceeds reversibly in solutions of pH > 9; hence under those conditions hydrogen peroxide yields a reversible anodic wave due to its oxidation to oxygen (cf. Eq. 1)

$$H_2O_2 \leftrightarrow O_2 + 2\,H^+ + 2\,e^- \qquad (12)$$

This oxidation wave can serve as a measure of hydrogen peroxide concentration, e.g., in studies of photosensitized autooxidations in solution in which hydrogen peroxide is one of the reaction products.

We made use of it [29] in research on photosensitized autooxidation of methanol in strongly alkaline media - a special kind of oxidation in which molecular hydrogen is produced besides hydrogen peroxide anion

$$2\,CH_3O^- + 2\,O_2 + OH^- \underset{sens}{\overset{h\nu}{\leftrightarrow}} 2\,HCOO^- + H_2 + HO_2^- + H_2O \qquad (13)$$

Measuring the anodic wave of hydrogen peroxide allowed us to follow the complex kinetics of this photoreaction under various experimental conditions and with various sensitizers. A direct irradiation of the dropping mercury electrode polarized in the potential region of the limiting current of H_2O_2 started the photoreaction in which now one of the reactants - oxygen - was being generated from the reaction product - hydrogen peroxide - according to reaction (12) at the electrode. This catalytic effect resulted in an instantaneous strong increase of the anodic current which fell back to its "dark" value at the moment the light was turned off. This kind of polarographic photoeffect is an example how polarography can be useful in research on the photodynamic effect which is currently studied for its medical applications [30].

1.3 Electroreduction of oxygen to water

For its reduction to two molecules of water one oxygen molecule needs four electrons and four protons :

$$O_2 + 4\,e^- + 4\,H^+ \rightarrow 2\,H_2O \qquad (14)$$

The first two electrons and protons can be accepted easily and quickly, as the bond between the two oxygen atoms remains preserved. However, accepting the next two electrons and protons is combined with breaking of this bond. On majority of metallic electrodes and in most solutions, the second stage of oxygen reduction is a slow irreversible process. Only activated surfaces of Pt or Ag accelerate this stage, so that on them oxygen reduction occurs in one four-electron step.

On mercury, the difference between the half-wave potentials of the first and second oxygen reduction waves amounts to 0.8 V. That allows, e.g., an independent determination of hydrogen peroxide in the presence of oxygen, as the reduction wave of H_2O_2 simply adds to the first reduction wave of oxygen.

In the metabolic processes of living systems, an accumulation of hydrogen peroxide as intermediate of oxygen reduction would have noxious effects because hydrogen peroxide, which can pass through cell membranes, acts as a source of strongly reactive OH radicals prone to attack living tissues. To prevent this danger, the organisms have developed protective reaction systems which decompose H_2O_2. By means of polarography and voltammetry models of such systems can be conveniently studied [5-7].

1.3.1 Catalysis by iron compounds - one of the key reactions of life

The great potential gap between the first and the second reduction step of oxygen on mercury electrodes can be narrowed by an addition of a catalyst into the solution which helps to split the O-O bond in hydrogen peroxide. Many such compounds have been already studied by polarography [31]; as a rule, the catalyst adsorbs at the electrode surface and by a specific mechanism lowers the overvoltage of H_2O_2 electroreduction. The extent of the shift of the second oxygen wave towards the first one is a measure of efficiency of each particular catalytic action. Most efficient of these catalysts are iron-containing compounds; of those only catalase - protein containing four ferri-protoporphyrins in molecule - was known to change the two-stage polarographic oxygen reduction into one four-electron process [7].

Recently we have found [32] that in aqueous solutions containing simple ferrous ions in concentrations higher than 0.1 mol dm^{-3} oxygen is also reduced in a single polarographic four-electron wave or voltam-metric peak corresponding to the uptake of four electrons (***Fig. 7***). A detailed research has shown that the catalytically active solutions contain a small concentration (about 10^{-5} mol dm^{-3}) of a bi- or polynuclear hydroxo complex of iron which behaves as a reversible redox system at the electrode with its potential in the region of the ascending part of the first oxygen wave (***Fig. 8***).

While catalase accelerates the disproportionation of hydrogen peroxide

$$2\,H_2O_2 \rightarrow O_2 + 2\,H_2O \qquad (15)$$

the ferro-ferri hydroxo complex achieves the same catalytic effect by mediating electron transfer between hydrogen peroxide and the electrode and thus accelerating the complete electroreduction of oxygen:

$$[Fe(II)X]^{n-} + H_2O_2 \rightarrow [Fe(III)X]^{(n-1)-} + OH^- + OH \text{ (solution)} \quad (16)$$

$$OH + e^- \rightarrow OH^- \text{ (electrode)} \quad (17)$$

$$[Fe(III)X]^{(n-1)-} \rightarrow [Fe(II)X]^{n-} - e^- \text{ (electrode)} \quad (18)$$

$$2\,OH^- + 2\,H^+ \rightarrow 2\,H_2O \text{ (solution)} \quad (19)$$

$$H_2O_2 + 2\,e^- + 2\,H^+ \rightarrow 2\,H_2O \quad (20)$$

The redox reaction (16) apparently occurs as an inner-sphere electron transfer with H_2O_2 entering as a ligand into the coordination sphere of the ferro-hydroxo complex.

Our results have shown that the decomposition of the undesirable hydrogen peroxide can be catalyzed not only by enzymes but also by an inorganic hydrolytic iron complex. A protective action by such a mechanism has presumably played an important role in the primary stages of development of life on earth when the simple ancestors of algae were beginning to photosynthesize in the shallow warm sea lagoons above the rich deposits of iron ores.

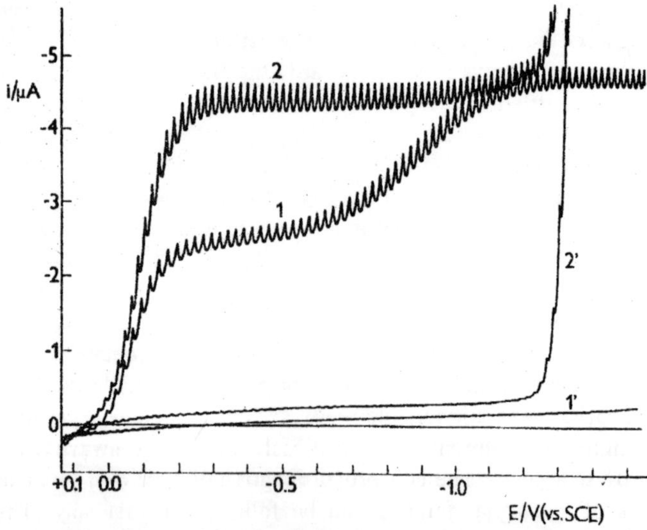

Fig. 7: **DC polarographic reduction of air oxygen in aqueous solutions**
1: 0.2M $(NH_4)_2SO_4$ and 0.2M $MgSO_4$; 2: 0.2M $(NH_4)_2Fe(SO_4)_2$. Curves 1' and 2' in deaerated solutions

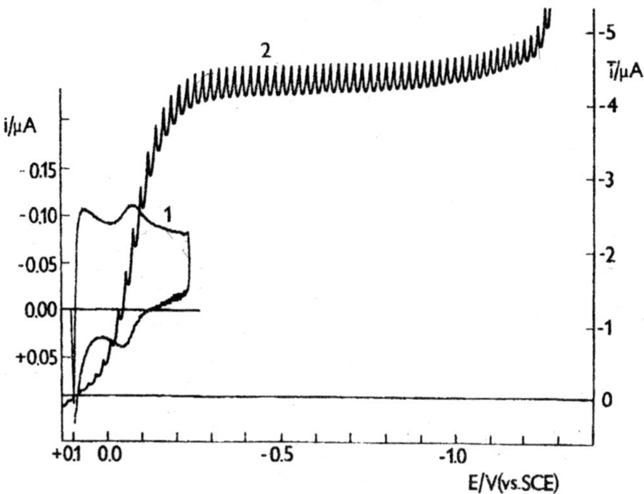

Fig. 8: **Comparison of the catalyst and its effect**

Curve 1: Cyclic voltammogram showing the reversible redox system in a deaerated 0.2M solution of $(NH_4)_2Fe(SO_4)_2$. HMDE, scan rate 50 mV s^{-1}. Current scale on the left. Curve 2: DC polarogram of the same solution after saturation with air oxygen. Current scale on the right

1.4 Electrochemical measurements of oxygen concentration

Since the appearance of the first papers on polarographic reduction of oxygen [1-4] until the nineteen fifties, many authors were using the dropping mercury electrode for determination of oxygen in various media including biological ones [12]. As a measure of oxygen concentration served usually the height of the limiting current corresponding to the total four-electron reduction of oxygen to water. By measurement at a fixed potential from the range of the plateau of the second oxygen wave it was possible to determine oxygen in a single measurement or continuously to ± 1 % at pressures ranging from 10^{-6} to 10^{-9} Pa. The content of oxygen in single breaths or in minute volumes of biological fluids could be followed in that way. The constant renewal of the electrode surface by spontaneous dropping provided perfectly reproducible conditions for oxygen measurements.

However, the operations with dropping mercury electrode were time-consuming and required special skill. The mercury constantly accumulating by dropping at the bottom of the cell was a nuisance in practical applications and needed sophisticated cell designs. Moreover, mercury has been considered as potentially toxic for organic systems. The latter problem could be partly solved by separating the mercury surface from the measured solution by a semi-permeable membrane, e.g. by cellophane (***Fig. 9***) [9]. The progress in the knowledge of behaviour of solid electrodes and in electronic instrumentation finally led to replacement of the dropping mercury electrode for practical oxygen measurements by specially designed solid (mostly Pt) electrodes, forming part of the so-called Clark sensor.

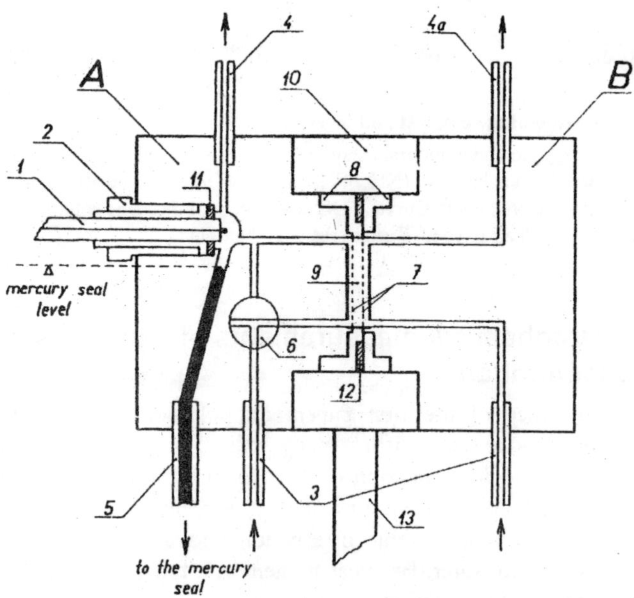

Fig. 9: **Scheme of the analyzer of oxygen consumption by tissue sections**
1 - dropping mercury electrode; 2 - fixing screw; 3 - solution inlets; 4 - silver tube serving as reference electrode and as solution outlet from part A of the cell; 4a - solution outlet from part B; 5 - mercury outlet; 6 - tap; 7 - cellophane membranes; 8 - retainer rings for fixing the membrane; 9 - space for the tissue under investigation; 10 - ring joining parts A and B of the cell; 11, 12 - gaskets; 13 - metal rod

1.4.1 The Clark sensor and its modifications

In the original Clark sensor constructed in 1953 [33], the working electrode was made of platinum, silver wire was used as the reference electrode and both the electrodes were dipped in a small volume of saturated potassium chloride solution and covered by a polyethylene film permeable to oxygen. In this way, the entire electrochemical system was enclosed in one tube which was dipped in the measured liquid and the leads from the electrodes were connected to a measuring apparatus, e.g., to a polarograph. The role of the polyethylene film (or of equally acceptable similar membranes, e.g., silicone rubber or cellophane) was, firstly, to prevent blocking or poisoning of the electrodes by adsorption or deposition of organic matter from the biological media on their surface which would render the measurements irreproducible, and, secondly, to eliminate the effect of uncontrolled convection on the electrolytic current when measuring in stirred or flowing liquids. One drawback of the Clark sensor has been its sensitivity towards temperature changes basically due to the high temperature coefficient of the membrane permeability for oxygen.

In course of measurement with the Clark sensor, the oxygen concentration in the small space between the electrode surface and the membrane is gradually exhausted and in order to repeat or to continue the measurement the sensor has to be allowed to regenerate its initial condition. For this reason it is advisable to polarize the sensor only in short intervals. In our laboratory we developed a method of oxygen measurements by short voltage pulses applied to the sensor after each 10 s [34]. The requirement of satisfactory reproducibility was met by insertion of a small spacer between the electrode and the membrane so that the diffusion conditions for oxygen would be exactly defined, fixed and maintained (***Fig. 10***).

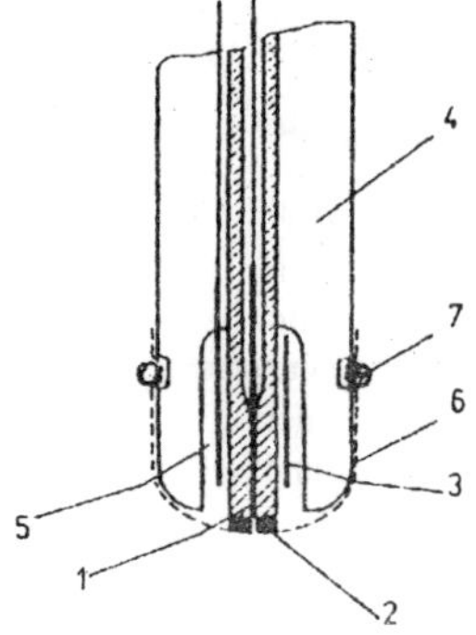

Fig. 10: **Schematic representation of a modified Clark-type sensor for pulse measurements**

1 - indicator electrode; 2 - spacer; 3 - reference electrode; 4 - sensor body; 5 - electrolyte space; 6 - covering membrane; 7 - O-ring fixing the membrane

For measurements of oxygen uptake or production in biological media it is essential that the sensor the studied system as little as possible by its own consumption of oxygen through electroreduction affects. With this aim we have developed miniature sensors with surface area of the working electrode less than 0.01 mm^2 which can be, moreover, sterilized at 120 °C. For measuring oxygen uptake by a small chicken embryo 40 hours after incubation it was necessary to modify the Clark sensor in the way that its tip was divided into seven symmetrically spaced indicator electrodes 0.05 mm in diameter, 0.2 mm apart, with a common reference electrode (***Fig. 11***) [35]. The arrangement works as seven independent miniature sensors indicating the different rates of oxygen uptake by different parts of the embryo.

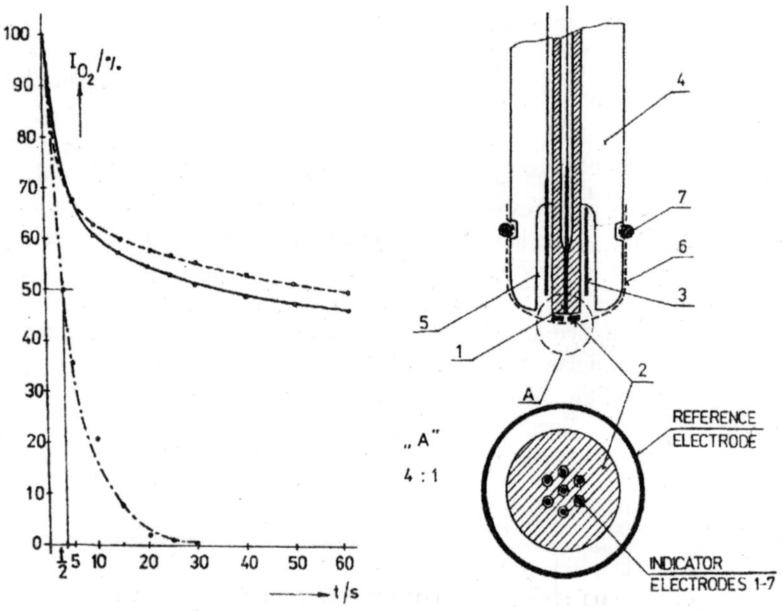

Fig. 11: **a) Example of oxygen consumption curves**

- . - . - control; ——— experimental; - - - - - out of the area.

b) Schematic diagram of the modified oxygen sensor

1 - array of seven indicator electrodes; 2 - spacer; 3 - reference electrode; 4 - body of the sensor; 5 - space filled with the electrolyte; 6 - covering membrane; 7 - fixing ring

1.4.2 Biological and medical applications

The practical applications of the Clark sensor are innumerable. In our laboratory we developed a special apparatus for the study of photosynthetic activity of water plants in the Institute of Botany, Academy of Sciences of the Czech Republic, in Třeboň. The apparatus allowed instantaneous or continuous measurement and recording of oxygen concentration, of pH and of temperature and provided automatic correction of the measured values for temperature changes [36].

We have also contributed to utilization of electrochemical oxygen measurements in medical practice [37]. In collaboration with several clinical laboratories, we designed special Clark sensors for:

a) in situ measurement of oxygen uptake by the tongue epithel (it was found to differ substantially on the upper and on the lower part of the tongue);
b) comparison of the respiratory activity of rabbit aorta in experimental atherosclerosis with the activity of a normal aorta;
c) study of oxygen uptake by human skin in various parts of the body surface and under various conditions;
d) determination of oxygen in air exhaled by patients and in gas mixtures used in inhalant narcotization;
e) oxygen measurement in blood with the sensor mounted directly in the piston of an injection syringe.

Great care was taken in developing a special measuring technique for studying oxygen uptake by chicken embryos [35]. This research was aimed at an extensive survey of noxious effects which a polluted environment can exert upon occurrence of inborn defects of living species acquired during the foetal development.

1.5 Conclusions

The survey of our results was intended to demonstrate that the electroreduction of oxygen, although studied since many years in laboratories all over the world, still continues providing new topics for basic and applied research. Due to the role oxygen plays on earth, the data on the mechanisms of the electrode processes of oxygen and its reduction products obtained with mercury electrodes, as well as the information on oxygen uptake or production in biological processes obtained with Clark sensors, contribute towards better understanding of the phenomenon of life and of the ways how to preserve it.

1.6 References

1. Heyrovský J.: Časopis českosl. lékárnictva 7, 242 (1927).
2. Heyrovský J.: Arhiv Hem. Farm. 5, 163 (1931).

3. Vítek V.: Chim. Industrie *29,* 215 (1933).
4. Vítek V.: Collect. Czech. Chem. Commun. *7*, 537 (1935).
5. Brdička R., Tropp C.: Biochem. Z. *289,* 301 (1937).
6. Brdička R., Wiesner K.: Collect. Czech. Chem. Commun. *12,* 39 (1947).
7. Koutecký J., Brdička R., Hanuš V.: Collect. Czech. Chem. Commun. *18,* 611 (1953).
8. Koryta J.: Collect. Czech. Chem. Commun. *18,* 21 (1953).
9. Šerák L.: *Advances in Polarography*, (Longmuir I.S., Ed.). Pergamon Press, London 1960. P. 1057.
10. Kůta J., Koryta J.: Collect. Czech. Chem. Commun. *30,* 4095 (1965).
11. Březina M.: Ber. Bunsenges. Phys. Chem. *77,* 869 (1973).
12. Šerák L.: in: *Electroanalytical Methods in Chemical and Environmental Analysis* (Kalvoda R., Ed.), Plenum Press, NewYork 1987. P. 174.
13. Heyrovský M., Vavřička S.: J. Electroanal. Chem. *332,* 309 (1992).
14. Rigo A., Tomat R., Rotilio G.: J. Electroanal. Chem. *57,* 291(1974);
Rigo A., Viglino P., Rotilio G.: Anal. Biochem. *68,* 1 (1975);
Rigo A., Rotilio G.: Anal. Biochem. *81,* 157 (1977).
15. Petersen R. A., Evans D. H.: J. Electroanal. Chem. *222,* 129 (1987).
16. Maricle D. L., Hodgson W. G.: Anal.Chem. *37,* 1562 (1965);
Peover M. E., White B. S.: Electrochim. Acta *11*, 1061 (1966);
Sawyer D. T., Roberts J. L., Jr.: J. Electroanal. Chem. *12*, 90 (1966).
17. Kastening B., Kazemifard G.: Ber. Bunsenges. Phys. Chem. *74*, 551 (1970);
Kastening B., Holleck L.: J. Electroanal. Chem. *27*, 355 (1970);
Chevalet J., Rouelle F., Gierst L., Lambert J.P.: J. Electroanal. Chem. *29*, 201 (1972);
Divíšek J., Kastening B.: J. Electroanal. Chem. *65*, 603 (1975).
18. Fedurco M., Šestáková I., Veselá V.: Langmuir, in press.
19. Harrison I. P.: *Metalloproteins*. Verlag Chemie, Weinheim 1985.
20. Florence T. M.: Biochem. J. *189*, 507 (1980).
21. Fedurco M.: Thesis, P. J. Šafárik University, Košice 1988.
22. Heyrovský M., Mader P., Veselá V., Fedurco M.: J. Electroanal. Chem. *369*, 53 (1994).
23. Koutecký J.: Collect. Czech. Chem. Commun. *18*, 311 (1953).
24. Fedurco M., Veselá V., Markušová K., Šipulová A.: Extended Abstracts, 14th International Congress of Biochemistry, Prague, July 10-15, 1988.
25. Fedurco M., Veselá V., Heyrovský M.: Electroanalysis *5*, 257 (1993).
26. Fedurco M., Veselá V., Heyrovský M., unpublished results.
27. Kosaka H., Imaizumi K., Tyuma I.: Biochim.Biophys.Acta *581*, 184 (1979).
28. Srivastava S. K. (Ed.): *Red Blood Cell and Lens Metabolism*. Elsevier, North Holland, 1980. PP. 123-137.
29. Heyrovský M., Shumov Yu. S.: Collect. Czech. Chem. Commun. *41*, 1860 (1976).
30. Kessel D.: Photochem. Photobiol. *39*, 851 (1984).
31. Březina M.: Collect. Czech. Chem. Commun. *22,* 339 (1957).
32. Heyrovský M., Vavřička S.: J. Electroanal. Chem. *353*, 335 (1993).
33. Clark L.C., Jr., Wolf R., Granger D., Taylor Z.: J. Appl. Physiol. *6*, 189 (1953).
34. Šerák L.: J. Total Environ. *37*, 107 (1984).
35. Šerák L., Jelínek R., Hauser V.: J. Electroanal. Chem. *226*, 193 (1987).
36. Šerák L., Čáp J., Pokorný J.: Proc. J. Heyrovský Memorial Congress on Polarography, Prague 1980, Vol.II. P. 158.
37. Kreuzer F., Kimmich H. P., Březina M.: in: *Medical and Biological Applications of Electrochemical Devices* (Koryta J., Ed.). J. Wiley, New York 1980. P. 173.

2

ION VOLTAMMETRY AT THE INTERFACE BETWEEN TWO IMMISCIBLE ELECTROLYTE SOLUTIONS

Zdeněk Samec

Abstract

Our contributions to the study of the ion transfer between water and an organic solvent are summarized. Attention is focused on the application of ion voltammetry and related techniques in measurements of the simple and facilitated ion transfer. Electrolysis at the interface between two immiscible electrolyte solutions provides the basis for a direct determination of various ions in the environment, an insight into liquid extraction processes, and an alternative way of extracting an ion from one phase to another.

Key Words

immiscible electrolytes, simple and facilitated ion transfer, voltammetry, polarography, electrolysis

2.1 Introduction

Electrochemistry dealing with the structure and charge transfer at the interface between two immiscible electrolyte solutions (ITIES) has a long history dating back to the beginning of this century [1]. Modern experimental approach to ITIES was founded by Gavach and coworkers [2] in 70', while Koryta [3] introduced the key theoretical concept of polarizability of an ITIES at about the same time. Various aspects of electrochemistry at ITIES have been reviewed by the present author [4,5], as well as by others [6-10].

The primary purpose of this review is to summarize our contributions to the study of the ion transfer between water and an organic solvent by voltammetric techniques [11-38]. Most of experimental work has been devoted to the water/nitrobenzene (NB) [11,13,14,16-27,30-35,37] and water/1,2-dichloroethane (DCE) [20,27-29,33,35,36] interfaces. We have also been interested in the use of o-nitrophenyl octyl ether (o-NPOE) [38], which had been found to be a suitable plasticizer for liquid-membrane ion-selective electrodes (ISE) [39,40].

Two types of ion transfer have been investigated. First one is the so-called simple ion transfer, which can be described

$$X^{z_i} (w) \longleftrightarrow X^{z_i} (o) \quad (1)$$

where X^{z_i} is an ion with the charge number z_i, and w and o denote the aqueous and the organic solvent phase, respectively. The other type is represented by the ion transfer (1) coupled to a chemical reaction, such as ion association or the formation of a complex at the ITIES,

$$X^z (w) + Y^s (o) \longleftrightarrow XY^{z+s} (o) \quad (2)$$

where Y^s is a counter ion or a ligand with the charge number s. The overall reaction (2) is usually denoted as the facilitated ion transfer.

2.2 Results and Discussion

2.2.1 Basic Concepts

Consider a binary electrolyte BA, which dissociates into the cation B^+ and anion A^-. The equilibrium distribution of the electrolyte BA between two immiscible solvents, e.g. water (w) and an organic solvent (o), is characterized by the electrolyte partition coefficient K_{BA},

$$K_{BA} = a_{\pm}^{o}/a_{\pm}^{w} = \gamma_{\pm}^{o} c_{BA}^{o} / \gamma_{\pm}^{w} c_{BA}^{w} = \exp(-\Delta_w^o G_{BA}^{\circ}/RT) \quad (3)$$

where R is the gas constant, T is the absolute temperature, $a_{\pm}^{s}$, $\gamma_{\pm}^{s}$ and c_{BA}^{s} are the mean activity, mean activity coefficient and the bulk concentration of BA in the phase s, respectively, and $\Delta_w^o G_{BA}^{\circ}$ is the standard Gibbs energy of transfer of BA from w to o. Since the mean activity of the electrolyte can be given in terms of the ion activities a^{+s} and a^{-s}, i.e. $a_{\pm}^{s} = (a^{+s}\ a^{-s})^{1/2}$, it is convenient to express the partition coefficient K_{BA} in terms of the single ion partition coefficients K_+ and K_-, i.e.

$$K_{BA} = (a^{+o}\ a^{-o})^{1/2}/(a^{+w}\ a^{-w})^{1/2} = (K_+K_-)^{1/2} \quad (4)$$

It can be written for a single-ion partition coefficient that

$$K_i = a_i^{o}/a_i^{w} = \exp(-\Delta_w^o G_i^{\circ}/RT) \quad (5)$$

where $\Delta_w^o G_i^{\circ}$ is the standard Gibbs energy of the ion transfer. Hence,

$$\Delta_w^o G_{BA}^{\circ} = (\Delta_w^o G_+^{\circ} + \Delta_w^o G_-^{\circ})/2 \quad (6)$$

From the electrochemical point of view, the partition of a charged species between the phases w and o necessarily gives rise to the electrical potential difference, $\Delta_w^o \phi = \phi(w) - \phi(o)$, which is given by a Nernst-type equation for the ion transfer [1],

$$\Delta_w^o \phi = \Delta_w^o \phi_i^{\circ} + (RT/z_iF) \ln (a_i^{o}/a_i^{w}) \quad (7)$$

where F is the Faraday constant and z_i is the ion charge number. The standard ion transfer potential $\Delta_w^o \phi_i^{\circ}$ is determined by the standard Gibbs energy of transfer of the ion i from the aqueous to the organic solvent phase, $\Delta_w^o G_i^{\circ}$,

$$\Delta_w^o \phi_i^o = - \Delta_w^o G_i^o / z_i F \tag{8}$$

While the single ion transfer energy $\Delta_w^o G_i^o$ is not an experimentally accessible quantity, the standard Gibbs energy of transfer of BA, $\Delta_w^o G_{BA}^o$, can be inferred from the partition or solubility measurements. Then, on the basis of a non-thermodynamic hypothesis [3], the single ion transfer functions can be determined. The most common TPAsTPB hypothesis [41] suggests that the single ion transfer energies of tetraphenylarsonium cation ($TPAs^+$) and tetraphenylborate anion (TPB^-) are equal for any couple of solvents.

Voltammetry of ITIES (VITIES) offers a unique method of measuring $\Delta_w^o \phi_i^o$ for an ion i relative to the standard ion transfer potential and the standard Gibbs energy of transfer of a reference ion. As shown by Koryta [3], a suitable system consists of two immiscible electrolyte solutions, such as

$$\underset{(w)}{X^{z_i}, B_1A_1} \quad || \quad \underset{(o)}{B_2A_2} \tag{9}$$

where X^{z_i} is the ion studied, and B_1A_1 and B_2A_2 are base electrolytes in the phase w and o, respectively. The standard Gibbs energies $\Delta_w^o G_i^o$ for ions B_1^+ and A_1^- must be large and positive and those for ions B_2^+ and A_2^- must be large and negative. Then the inequality can be fulfilled,

$$\max(\Delta_w^o \phi_{B2}^o, \Delta_w^o \phi_{A1}^o) << \Delta_w^o \phi_X^o << \min(\Delta_w^o \phi_{B1}^o, \Delta_w^o \phi_{A2}^o) \tag{10}$$

which ensures that the ion transfer (1) can be measured without interference from the transfer of the base electrolyte ions. In the absence of the ion X^{z_i}, there is a range of the potential differences $\Delta_w^o \phi$

$$\max(\Delta_w^o \phi_{B2}^o, \Delta_w^o \phi_{A1}^o) << \Delta_w^o \phi << \min(\Delta_w^o \phi_{B1}^o, \Delta_w^o \phi_{A2}^o) \tag{11}$$

within which the ITIES can behave as an ideally polarizable interface.

2.2.2 Cyclic Voltammetry of Ion Transfer

The transport and thermodynamic parameters of the ion transfer reactions (1) and (2) can be most conveniently measured by cyclic voltammetry (CV). In the pioneering study [11], a four electrode system has been introduced, which comprises two reference electrodes for polarization of ITIES and two counter electrodes supplying the electrical current. ***Fig. 1*** shows a suitable configuration of the electrodes in the electrolytic cell, which can also be represented by the scheme

$$\underset{RE1}{Ag \mid AgX} \mid \underset{(w)}{BX} \,||\, \underset{(o)}{RY} \mid \underset{(w')}{RX} \mid \underset{RE2}{AgX \mid Ag'} \tag{12}$$

where RE1 is a silver/silver halide reference electrode, RE2 is a reference electrode, which is ion-selective to the cation R^+ of the organic phase o, the aqueous electrolyte BX is, e.g., LiCl and the organic electrolyte RY is, e.g., tetrabutylammonium tetraphenylborate. The potential E of the cell (12) is then controlled or measured in a potentiostatic or galvanostatic experiment, respectively. An important detail has soon been recognized [14] that it is necessary to compensate for the ohmic potential drop between the tips of the reference electrode Luggin capillaries. A simple scheme of a four-electrode potentiostat with positive feedback ohmic drop compensation is given in ***Fig. 2***.

A cyclic voltammogram for ion transfer across ITIES has the shape of a peak known from the classical voltammetry of a simple electron transfer at a planar electrode. The voltammogram is characterized by the peak current I_p, the peak potential E_p and the peak potential difference ΔE_p between the anodic (positive current) and cathodic (negative current) peaks. Since the transport problems are quite analogous [15], the Nicholson-Shain theory of stationary electrode polarography [42] has been used to evaluate the ion diffusion coefficient D and the standard potential E° (i.e. the standard potential difference, $\Delta_w^{\,o}\phi_i^{\circ}$) from experimental voltammetric data. Provided that the ion transfer (1) is controlled by diffusion, i.e. $E_p = (59/z)$ mV,

$$I_p = \pm\,(2.69 \times 10^5)\; z^{3/2} A\, D^{1/2} v^{1/2} c^{o} \;\text{(in A)} \tag{13}$$

where A is the interfacial area (cm^2), v is the polarization rate (V s^{-1}) and c° is the bulk ion concentration (mol cm^{-3}), the reversible half-wave potential being given by $E_{1/2}^{\text{rev}}$

$$E_{1/2}^{\text{rev}} = E^{\circ} + (RT/2zF)\ln(D^{w}/D^{o}) \tag{14}$$

which should precede the peak potential by (28.5/z) mV. A cyclic voltammogram of the tetraethylammonium ion transfer across the water/ o-NPOE interface is illustrated in ***Fig. 3***.

Cyclic voltammetry can be combined with the convolution analysis [15], which makes it also possible to infer the kinetic parameters for the interfacial ion transfer [12] from CV data. In this approach, which is known as the convolution potential sweep voltammetry (CPSV) [43], the convolution current $m(t)$ is calculated as a function of time t from the electrical current $I(t)$,

$$m(t) = \pi^{-1/2} \int_0^t I(t)\,(t-\tau)^{-1/2}\, d\tau \tag{15}$$

where τ is the integration dummy variable. For ion transfer (1), the convolution current m is a function of the potential E, i.e. the potential difference $\Delta_w^{\,o}\phi$, [15],

$$(m_d - m)m^{-1} = [I(D^{w})^{1/2}/m\,k_f] + (D^{w}/D^{o})^{1/2}\exp[-zF(E-E^{\circ})/RT] \tag{16}$$

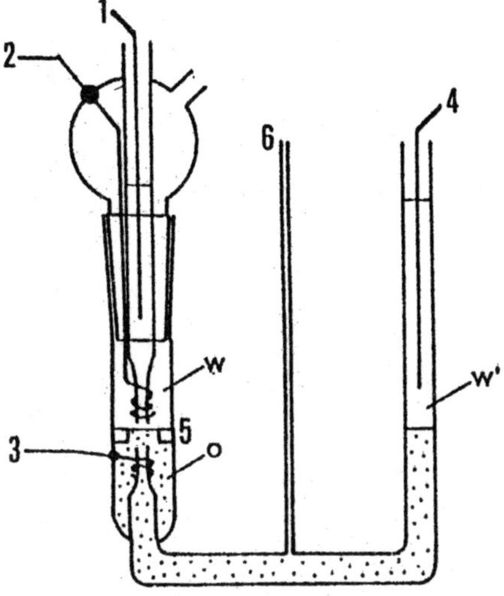

Fig. 1: **Scheme of the four-electrode cell with the aqueous (w, w') and the organic solvent (o) phases**

1, 4 - Ag/AgCl reference electrodes, 2, 3 - platinum counter electrodes, 5 - a glass barrier with the round hole for the liquid/liquid interface, 6 - a tube connected to a syringe for adjustment of the interface. After [24].

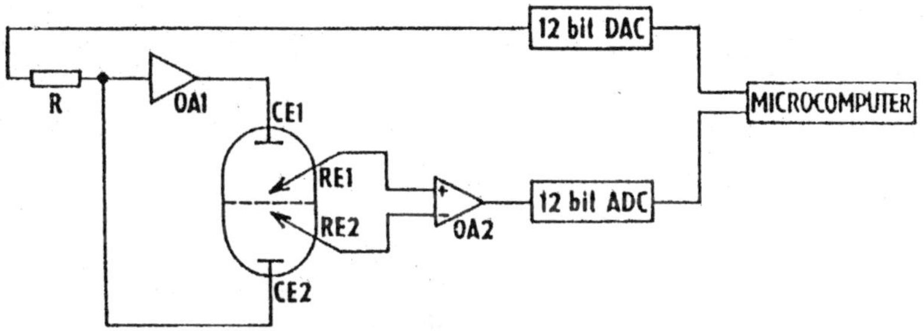

Fig. 2: **Scheme of the four-electrode potentiostat**

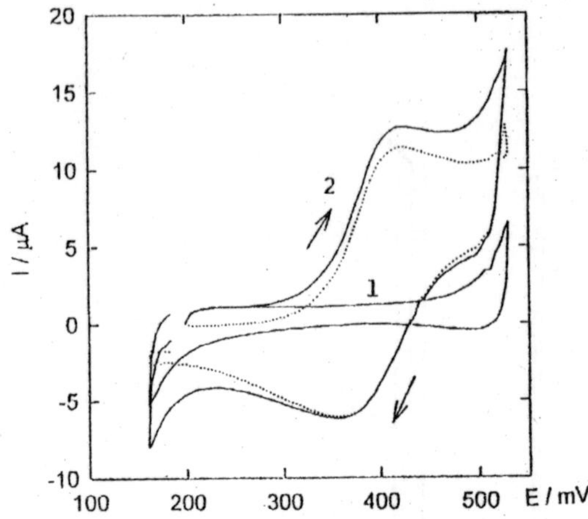

Fig. 3: **CV of 0.01M LiCl in water and 0.02M tetrapentylammonium tetraphenylborate in o-NPOE in the absence (1) and in the presence (2) of tetraethylammonium ion**

Concentration of tetraethylammonium ion 0.25 mmol l^{-1} and the sweep rate 0.1 V s^{-1}. Dotted line shows the current due to the transfer obtained by subtracting the two voltammograms. After [38].

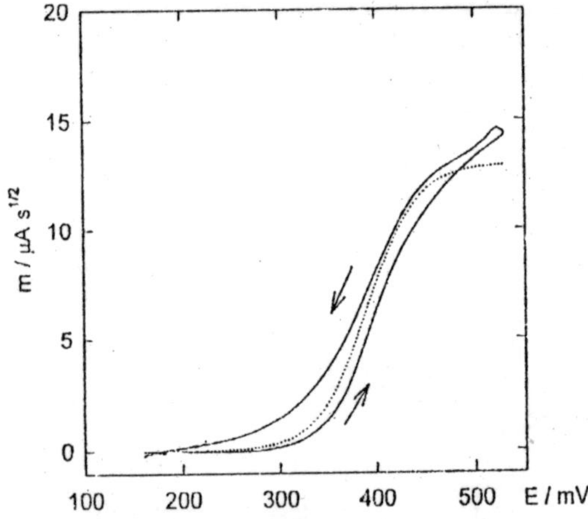

Fig. 4: **CPSV of tetraethylammonium ion transfer across the water/o-NPOE interface at a sweep rate of 0.1 V s^{-1} and an ion concentration of 0.25 mmol l^{-1}**

The dotted line represents the theoretical curve corresponding to the reversible ion transfer as calculated from eq. (16) for the diffusion coefficient and the reversible half-wave potential derived from CV. After [38].

where k_f is the rate constant of the ion transfer from w to o (e.g., in cm s^{-1}). When the ion transfer reaction is fast, i.e. $k_f \rightarrow \infty$, the first term on the right hand side of Eq. (16) can be neglected, and Eq. (16) describes a reversible CPSV wave with the half-wave potential given by Eq. (14). An example of the CPSV analysis is shown in ***Fig. 4***. On the other hand, when the ion transfer is slow, the potential dependent rate constant k_f can be determined [16]. CPSV makes it also possible to infer the transport, thermodynamic and kinetic data for more complex ion transfer reactions, such as the ion transfer facilitated by the formation of complexes of various stoichiometries [18].

By using cyclic voltammetry, the ion diffusion coefficients and the standard Gibbs energies of ion transfer have been determined for Cs^+ [14], picrate [30,31,35], perchlorate [31,35], tetraalkylammonium cations R_4N^+ (R = methyl, ethyl, propyl) [25,31,35], choline and acetylcholine [21], protonated amines (aniline, benzylamine, 2-phenylethylamine, tyramine, 3-hydroxy-tyramine, noradrenaline, metanephrine, phenylephrine) [23] and viologen-dications R_2V^{2+} (R = methyl, ethyl, propyl, butyl, pentyl, heptyl or benzyl) [27,31] at the water/nitrobenzene interface, and alkali metal cations [29,36], chloride, bromide and iodide [36], laurylsulphate [28], picrate [35], perchlorate [35], tetraalkylammonium cations R_4N^+ (R = methyl, ethyl) [35], viologen-dications R_2V^{2+} (R = methyl, ethyl, propyl, butyl, pentyl, heptyl or benzyl) [27] at the water/1,2-dichloroethane interface.

Only a few CPSV kinetic analyses have been performed for the simple ion transfer [16,21,25]. Standard rate constants, i.e. the values of the rate constant k_f at the standard potential $E°$, are in a range of 0.02 - 0.05 cm s^{-1} [16,21,25]. However, evaluation of ion transfer rates from measurements involving potential sweeps or steps over wide potential ranges can be uncertain. This is mainly caused by the considerable potential potential drops, which must be compensated for. Moreover, mechanical instability of the ITIES arising from the potential dependence of the surface tension can influence the ion transport. Indeed, impedance measurements at an equilibrium potential difference yield somewhat higher standard rate constants, around 0.1 cm s^{-1} [30,31].

Selective complexation of biogenic amines by macrocyclic polyethers at the water/nitrobenzene interface was investigated by cyclic voltammetry [34]. The structure of the complex is shown in ***Fig. 5***. In the presence of dibenzo-18-crown-6, the transfer of β-phenylethylammonium ions (i.e. protonated β-phenylethylamine, tyramine, dopamine, noradrenaline, homoveratrylamine, amphetamine, phentermine, phenylephrine, ephedrine or metanephrine) is facilitated, so that the voltammetric peak of the ion transfer shifts toward negative potentials, cf. ***Fig. 6***. The shift in the half-wave potential,

$E_{1/2}$, is related to the equilibrium constant for the complexation in the organic phase, K^o,

$$E_{1/2} = -(RT/F) \ln [1 + K^o(\gamma^o)^2 c_L{}^o (D_{ML}{}^o/D_M{}^o)^{1/2}] \qquad (17)$$

where γ^o is the single ion activity coefficient, $c_L{}^o$ is the ligand concentration, $D_{ML}{}^o$ and $D_M{}^o$ are the diffusion coefficients of the complex and the ion, respectively, all in the organic solvent phase. The temperature dependence of the equilibrium constant was then examined and the entropy and enthalpy of complexation derived. The data confirm that selective complexation arises both from discrimination in binding these ions to the crown-ether macrocycles and from resolvation effects which modify the ion transfer energetics. These effects are closely related to the structure of the transferred ion and thus a chemical substitution (e.g. introduction of OH or CH_3 groups into the substrate ion molecule) can reverse the ion exchange equilibrium at the interface. The results throw some light on the regulation of the passage of polar molecules through the lipidic barriers of biological membranes.

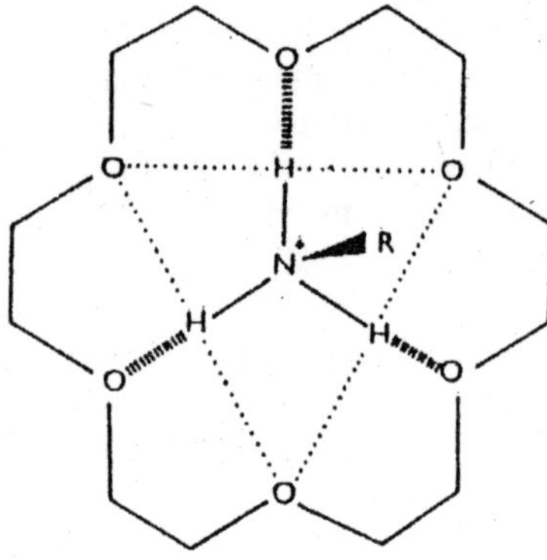

Fig. 5: **Structure of the complex formed between 18-crown-6 and a primary ammonium cation**

Facilitated transfer of proton, alkali- and alkaline earth-metal cations across the water/nitrobenzene interface was investigated [18] in the presence of acyclic and cyclic polyetherdiamides: N,N'-di[(11'-ethoxycarbonyl)undecyl]-N,N',4,5-tetra-methyl-3,6-dioxaoctanediamide (DODA) and 7,19-dibenzyl-2,3-dimethyl-7,19-diaza-1,4,10,13,16-pentaoxa-cyclo-heneicosane-6,20-dione (PEDA) in the nitrobenzene phase. These neutral synthetic compounds act as the ion carriers which mediate the ion transfer across the ITIES through the complex formation. By using CPSV, the metal-to-ligand stoichiometry $(r : s)$

and the thermodynamic, transport and kinetic parameters were evaluated for the single step charge transfer model, i.e. $r\ M^{z+}(w) + s\ L(o) = M_rL_s^{rz+}(o)$, where M^{z+} is the metal cation and L is the neutral ligand. The formation of 1 : 1 and 1 : 2 complexes was observed for monovalent and divalent cations, respectively, and the stability constants of these complexes were determined. While the facilitated transfer of monovalent cations was too fast to be determined, the kinetic parameters for the divalent ion transfer were obtained, which obviously reflect the effects of both the electrical double layer and the ligand structure or solvation [18].

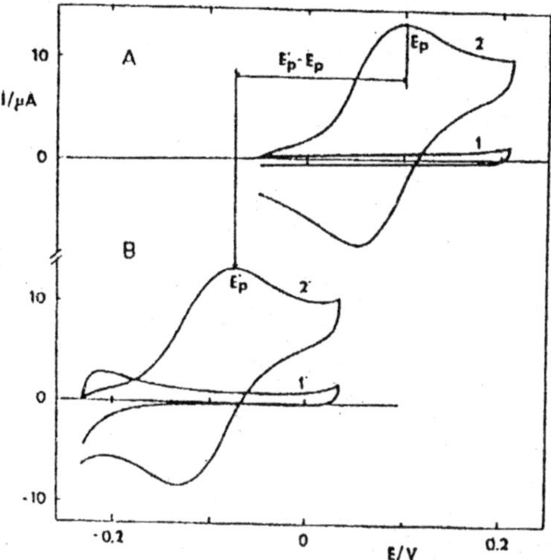

Fig. 6: **CV of base electrolytes (1, 1') and 1 mmol l⁻¹ phenylethylammonium ion (2, 2') in the absence (A) and presence (B) of 0.01M dibenzo-18-crown-6 in nitrobenzene**

Base electrolytes: 0.01M HCl in water and 0.01M tetraphenylarsonium dicarbollylcobaltate in nitrobenzene, sweep rate 0.02 V s^{-1}. After [34].

2.2.3 Polarography of Ion Transfer

Polarography with an electrolyte dropping electrode (EDE) was pioneered by Koryta [3]. In the further development, we have designed a four-electrode assembly [13], which was employed without substantial modifications by various authors to study the electrical double layer [44] or the ion transport [45] at ITIES. Fundamental factors in polarography with an EDE were examined [46] for both possible configurations, i.e. for the electrolyte solution dropping upward (i.e. an ascending electrolyte electrode), and for the electrolyte solution dropping downward. In addition to the classical potential-scan polarography [3,13], the current-scan polarography has been introduced [47]. The advantage of this method is an easy compensation of the ohmic potential drop. On the other hand, the enormous current density at the very beginning of the drop life can cause irreversible changes in the boundary state and conditions.

More recently [33], we developed a simple three-electrode polarographic assembly (***Fig. 7***) with an EDE which can be polarized by using a conventional three-electrode polarograph permitting ohmic drop compensation. The aqueous phase was dropping upward into the organic solvent phase from a PTFE capillary (0.02 cm in inner diameter, with a flow rate of 1-2 mg s^{-1} and the drop time of about 5 s). The electrolyte flow rate was controlled by the height of the aqueous electrolyte solution column, which was cca 40 cm to the capillary orifice. The reference and counter electrode couple for the aqueous phase in the original four-electrode assembly with an EDE [13] was replaced by a large-area (about 5 cm^2) Ag/AgCl reference electrode. The electrode was connected to the working electrode input of the three-electrode potentiostat, i.e. to the input of the current follower, which is maintained at the virtual ground. Polarization of this electrode was negligible, owing to low current densities involved. The reference electrode for the organic solvent phase was of the same type as in the scheme (12). A polarogram of base electrolytes in the absence and presence of tetramethylammonium ion is given in ***Fig. 8***. The polarographic method was then used to study the mechanism and thermodynamics of the facilitated transfer of the alkali metal cations from water to nitrobenzene or 1,2-dichloroethane in the presence of dibenzo-18-crown-6, dibenzo-24-crown-8 or dibenzo-30-crown-10 [33].

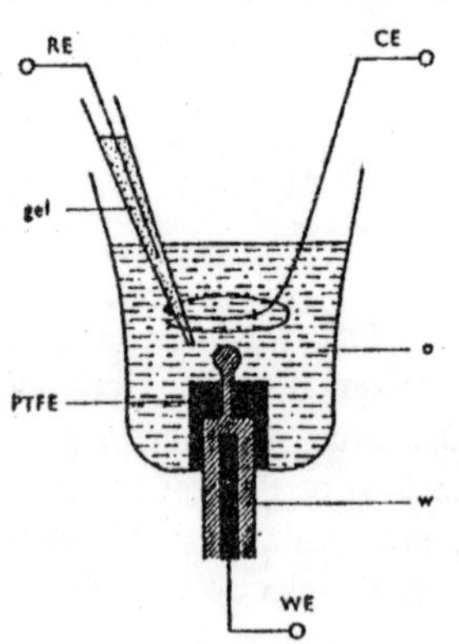

Fig. 7: **Three-electrode polarographic cell with the electrolyte dropping electrode**

WE - large-area Ag/AgCl working electrode, RE - Ag/AgCl reference electrode, CE - Pt wire counter electrode. After [33].

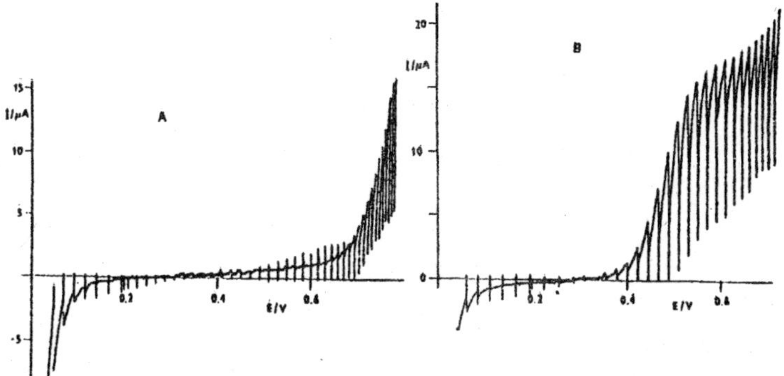

Fig. 8: **An EDE polarogram of the base electrolytes in the absence (A) and presence (B) of 0.825 mmol l^{-1} tetramethylammonium bromide in the aqueous phase**

Base electrolytes: 0.01M LiCl + 0.5M $MgSO_4$ in water and 0.01M tetraphenyl-arsonium dicarbollylcobaltate in 1,2-dichloroethane. After [33].

2.2.4 Analytical Applications

A three-electrode system with a hanging electrolyte drop electrode (HEDE) was designed for analytical exploitation of VITIES [17,19]. The assembly for the HEDE is shown in ***Fig. 9***. A small drop of the nitrobenzene solution (a) is formed at the capillary tip (b) of a glass tube, which is immersed in the test aqueous solution. This tip of about 1 mm i. d. is drawn from a glass tube of 4 mm i. d. The volume of the nitrobenzene drop is controlled by a calibrated microsyringe, which is connected to one inlet (b) of the glass tube. The metal wire counter electrode CE2 and the Ag/AgCl reference electrode RE2 are fixed in the other two outlets (e, f). The reference electrode is dipped in the reference aqueous solution, cf. the scheme (12). The counter electrode CE2 is a platinum or copper wire (h) insulated by Teflon (i), so that only a metal disc with an area of 0.126 mm^2 is exposed to the nitrobenzene solution (a). A motor-driven glass stirrer (j) and a large-area Ag/AgCl reference electrode RE1 are dipped in the aqueous test solution (c). The latter electrode also serves as a counter electrode, familiar to the above three-electrode polarographic assembly (***Fig. 7***).

The HEDE was used in determinations of micromolar concentrations of various ions in water, such as acetylcholine [17], calcium, barium and strontium [22], aniline, *p*-aminophenol, benzylamine and protonated catecholamines [23], nitrate, perchlorate and iodide [26] by differential pulse stripping voltammetry (DPSV). Methodology and the comparison with the theory of DPSV have been discussed in detail [19].

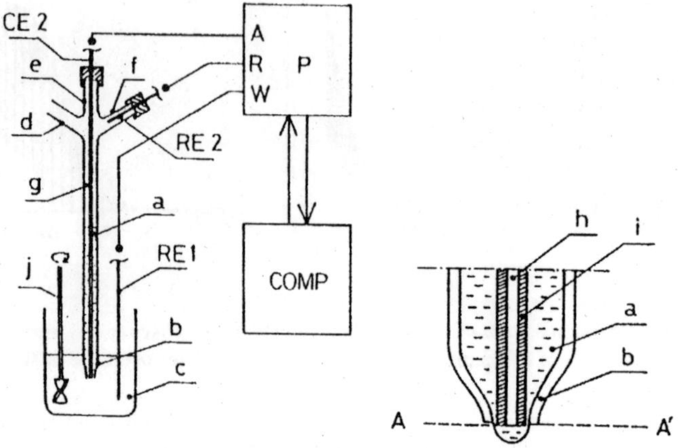

Fig. 9: **Assembly for the HEDE**

RE1, RE2 - Ag/AgCl electrodes, CE2 - counter electrode, P - a conventional three-electrode potentiostat, COMP - microcomputer. The right hand diagram shows the HEDE in greater detail (see the text). After [19].

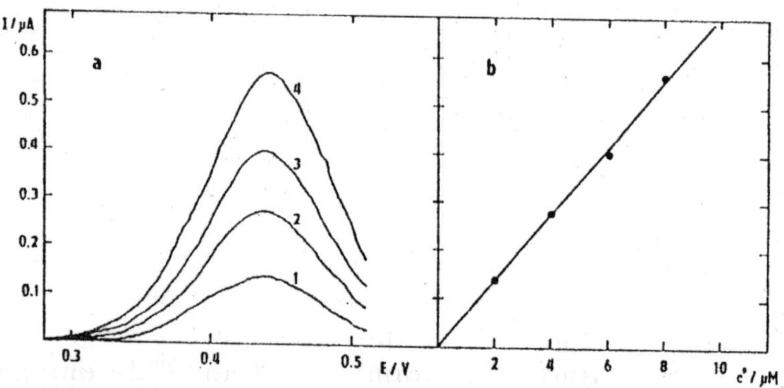

Fig. 10: **The DPSV of protonated aniline at a nitrobenzene HEDE**

Concentration of aniline in the aqueous phase (µmol l^{-1}): 2 (1), 4 (2), 6 (3) and 8 (4). Base electrolytes: 5mM HCl in water and 5mM tetrabutylammonium dicarbollylcobaltate in nitrobenzene. After [23].

Fig. 10 shows the DPSV of protonated aniline in water at various concentrations of aniline [23]. First, a stepwise voltage pulse (1mV/20ms) was applied to the HEDE starting at an initial potential of 0.190 V and increasing to a pre-electrolysis potential of 0.510 V. During the pre-electrolysis (60 s), the aqueous solution containing aniline was stirred (375 rpm). The stripping of aniline was accomplished by applying a series of staircase potential pulses (ΔE = 40 mV), with the pulse duration of 60 ms. It is seen from ***Fig. 10*** that the plot of the DPSV peak *vs.* the ion concentration exhibits a good linearity. The limit of determination of the method is at the 1 μmol l^{-1} level.

Electrical signals measured at ITIES often deteriorate due to instabilities of liquid/liquid interfaces, which have the origin in changes of the surface tension with the electrical potential difference, or are due to adsorption of the solution components at the interface. Therefore, attempts have been made to stabilize the ITIES by replacing one liquid phase with a polymer gel electrolyte, thereby forming a polymer gel electrolyte electrode (PGEE) [48]. The gel is prepared by mixing a suitable polymer (polyvinylchloride, agar) with one of the electrolyte solutions at an elevated temperature [49,50]. The mixture is then cooled to room temperature in a suitable mold. Such an electrode resembles solvent-polymer membrane ISE's, which have enjoyed broad analytical application [7]. Cyclic voltammetry and equilibrium impedance spectroscopy were used to study the transfer of tetraalkylammonium, choline, perchlorate and picrate ions across water/polyvinylchloride + nitrobenzene gel or agar + water gel/nitrobenzene boundaries, as a function of the polymer concentration [32]. Ion transfer rates reflect the change in the relative fraction of the electrolyte solution phase, which probably occupies submicrometer pores dispersed in the bulk or on the surface of the electrochemically inactive polymer matrix. Ion transfer across a polymer gel/liquid boundary is about as fast as that across a liquid/liquid boundary [32].

2.3 Conclusions

The results of this study can be relevant to environmental protection in two respects. First, voltammetry at ITIES (VITIES) is an amperometric analogue of potentiometry with liquid-membrane ISE's [7] and thus represents an alternative method for the determination of ions in the environment. Second, the VITIES can provide an insight into liquid extraction processes, e.g. in metal ion recovery or waste water treatment [40]. Furthermore, electrically driven ion transport across a solid supported liquid membrane offers an alternative way of extracting an ion from one phase to another.

2.4 References

1. Nernst W.: Z. Phys. Chem. *9*, 137 (1892).
2. Gavach C., Mlodnicka T., Guastalla J.: C. R. Acad. Sci. C. *266*, 1196 (1968).
3. Koryta J., Vanýsek P., Březina M.: J. Electroanal. Chem. Interfacial Electrochem., 211 (1977).
4. Samec Z.: Chem. Rev., 617 (1988).
5. Samec Z., Kakiuchi T.: Advances in Electrochemistry and Electrochemical Science, Gerischer H., Tobias C. W. (Eds.), Verlag Chemie, Weinheim, 1995.
6. Koryta J.: Ion-Select. Electrode Rev. *5*, 131 (1983).
7. Koryta J., Štulík K.: Ion Selective Electrodes, Cambridge University Press, Cambridge, 1983.
8. Kazarinov V. E. (Ed.): The Interface Structure and Electrochemical Processes at the Boundary Between Two Immiscible Electrolyte Solutions, Springer, Berlin, 1987.
9. Girault H. H., Schiffrin D. J.: Electroanalytical Chemistry, Vol. 15 (Bard A. J., Ed.), Marcel Dekker, New York, 1989. P.1.
10. Senda M., Kakiuchi T., Osakai T.: Electrochim. Acta *36*, 253 (1991).
11. Samec Z., Mareček V., Koryta J., Khalil M. V.: J. Electroanal. Chem. Interfacial Electrochem. *83*, 393 (1977).
12. Samec Z.: J. Electroanal. Chem. Interfacial Electrochem. *99*, 197 (1979).
13. Samec Z., Mareček V., Weber J., Homolka D.: J. Electroanal. Chem. Interfacial Electrochem. *99*, 385 (1979).
14. Samec Z., Mareček V., Weber J.: J. Electroanal. Chem. Interfacial Electrochem. *100*, 841 (1979).
15. Samec Z.: J. Electroanal. Chem. Interfacial Electrochem. *111*, 211 (1980).
16. Samec Z., Mareček V., Weber J., Homolka D.: J. Electroanal. Chem. Interfacial Electrochem. *126*, 105 (1981).
17. Mareček V., Samec Z.: Anal. Lett. *14* (B15), 1241 (1981).
18. Samec Z., Homolka D., Mareček V.: J. Electroanal. Chem. Interfacial Electrochem. *135*, 265 (1982).
19. Mareček V., Samec Z.: Anal. Chim. Acta *141*, 65 (1982).
20. Samec Z., Homolka D., Mareček V., Kavan L.: J. Electroanal. Chem. Interfacial Electrochem. *145*, 213 (1983).
21. Samec Z., Mareček V., Homolka D.: J. Electroanal. Chem. Interfacial Electrochem. *158*, 25 (1983).
22. Mareček V., Samec Z.: Anal. Chim. Acta *151*, 265 (1983).
23. Homolka D., Mareček V., Samec Z., Baše K., Wendt H.: J. Electroanal. Chem. Interfacial Electrochem. *163*,159 (1984).
24. Mareček V., Samec Z.: J. Electroanal. Chem. Interfacial Electrochem. *185*, 263 (1985).
25. Samec Z., Mareček V.: J. Electroanal. Chem. Interfacial Electrochem. *200*, 17 (1986).
26. Mareček V., Jänchenová H., Samec Z., Březina M.: Anal. Chim. Acta *185*, 359 (1986).
27. Hanzlík J., Samec Z.: Collect. Czech. Chem. Commun. *52*, 830 (1987).
28. Kontturi A. K., Kontturi K.: Acta Chem. Scand. *A42*, 192 (1988).

29. Samec Z., Mareček V., Colombini M. P.: J. Electroanal. Chem. Interfacial Electrochem. *257*, 147 (1988).
30. Wandlowski T., Mareček V., Samec Z.: J. Electroanal. Chem. Interfacial Electrochem. *242*, 291 (1988).
31. Wandlowski T., Mareček V., Holub K., Samec Z.: J. Phys. Chem. *93*, 8204 (1989).
32. Dvořák O., Mareček V., Samec Z.: J. Electroanal. Chem. Interfacial Electrochem. *284*, 205 (1990).
33. Samec Z., Papoff P.: Anal. Chem. *62*, 1010 (1990).
34. Dvořák O., Mareček V., Samec Z.: J. Electroanal. Chem. Interfacial Electrochem. *300*, 407 (1991).
35. Wandlowski T., Mareček V., Samec Z.: Electrochim. Acta *35*, 1173 (1990).
36. Sabela A., Mareček V., Samec Z., Fuoco R.: Electrochim. Acta *37*, 231 (1992).
37. Wandlowski T., Mareček V., Samec Z., Fuoco R.: J. Electroanal. Chem. Interfacial Electrochem. *331*, 765 (1992)
38. Samec Z., Trojánek A., Samcová E.: J. Electroanal. Chem. Interfacial Electrochem. (1992), in press.
39. Oesch U., Simon W.: Anal. Chem. *52*, 692 (1980).
40. Pletcher D., Walsh F. C.: Industrial Electrochemistry, 2nd Edition, Chapman and Hall, London, 1990.
41. Parker A. J.: Electrochim. Acta *21*, 671 (1976).
42. Nicholson R.S., Shain I.: Anal. Chem. *36*, 706 (1964).
43. Greness M., Oldham K. B.: Anal. Chem. *44*, 1121 (1972).
44. Kakiuchi T., Senda M.: Bull. Chem. Soc. Jpn. *56*, 1322 (1983).
45. Kihara S., Suzuki M., Sugiyama M., Matsui M.: J. Electroanal. Chem. Interfacial Electrochem. *249*, 109 (1988).
46. Kihara S., Suzuki M., Maeda K., Ogura K. Umetani S., Matsui M.: Anal. Chem. *58*, 2954 (1986).
47. Kihara S., Yoshida Z., Fujinaga T.: Bunseki Kagaku *31*, 297 (1982).
48. Osakai T., Kakutani T., Senda M.: Bunseki Kagaku *33*, E371 (1984).
49. Mareček V., Jänchenová H., Colombini M. P., Papoff P.: J. Electroanal. Chem. Interfacial Electrochem. *217*, 213 (1987).
50. Mareček V., Colombini M. P.: J. Electroanal. Chem. Interfacial Electrochem. *241*, 133 (1988).

3

PRINCIPLES OF INTERFACIAL MEASUREMENTS AND ADSORPTION VOLTAMMETRY WITH MERCURY ELECTRODES

Ladislav Novotný

Abstract

The aim of the present article is to provide basic orientation in the field of electrosorption methods and to give an account of the original contributions of the author and his co-workers respectively to this topic. Selected examples of actual solutions of the pertaining problems are given.

Key Words

adsorption voltammetry, polarography, electroanalysis, surfactants, interfacial activity, mercury electrodes and microelectrodes, determination of surfactants

3.1 Introduction

Adsorption and interfacial processes at electrodes constitute an integral part of electrode reactions. Their enhancement or suppression depends on the selection of the method and experimental conditions.

The methods based on adsorption accumulation of the test substance at the working electrode surface or the interface boundary [7-33, 56, 97-110] are gaining increasing importance in contemporary polarography, voltammetry and related techniques [1-6]. Adsorption stripping voltammetry (AdSV), or simply "adsorption voltammetry" is most frequently encountered in the present literature on electroanalysis. The foundations of this method were laid in the sixties [7, 9] and a considerable progress has recently occurred. The method is characterized by the use of stationary electrodes and by exact time monitoring of the adsorption accumulation of the test substance at the electrode/solution boundary; in the final stage a particular modification of the current-voltage curve is recorded. Measurements with quasi-stationary electrodes [11-14, 17] (i.e. polarographic measurements) and studies of time-dependent currents [19], shifts in the peak-potential [33], etc., are sometimes also included in this group of methods. The adsorption voltammetry has developed gradually in various modifications, at the beginning especially in combination with a.c. polarography (tensammetry) and a.c. oscillographic polarography, utilizing the principles of classical measurements of the double layer differential capacity and the electrocapillary curves. Its importance increased further with arrival of the sensitive methods of pulse polarography, especially of the differential pulse polarography and voltammetry.

The polarized mercury electrode has occupied an important position among the working electrodes in adsorption voltammetry, mainly because of its reproducibility and easy preparation. Until the eighties, polarography was almost exclusively based on the use of the dropping mercury electrode (DME), and this electrode was not suitable for further development of adsorption voltammetry. The main obstacle was the lack of a simple way of producing a reproducible stationary mercury drop. The then existing commercial micrometric system [5] for manual formation of hanging mercury drop electrodes (HMDE) yielded only 15-20 stationary drop electrodes per reservoir filling, reproducible within ± 2-5 %, which led to repeated use of one unrenewed, and hence "polluted" or at least "surface modified" drop and to a limited reproducibility of measurement.

A qualitative step forward in this respect was achieved at the beginning of the eighties by the introduction and commercial production of the static mercury drop electrode (SMDE) [12, 34-42] permitting the formation of tens to hundreds of thousands of renewed hanging (stationary) mercury drops, with a reproducibility of ± 1 % and better. It has brought about the programmable renewal of the HMDE which, as will be shown further, is one of the decisive factors in the study of the processes involved in adsorption accumulation. At present there is a miniaturized portable apparatus available for producing renewed mercury mini-, semimicro- and microelectrodes [12, 35, 37-39, 41, 43-51] with a wide range of optional electrode parameters.

The marked improvement in the measuring reproducibility, sensitivity and availability of the necessary equipment was accompanied by a remarkable progress in AdSV; the number of compounds that can be determined by AdSV is rapidly increasing. However, a routine application of polarographic and voltammetric principles with adsorptive accumulation can lead to irreproducible or useless results, as the signal results from adsorption-desorption processes which are essentially different from ordinary electrooxidations or electroreductions. It is thus highly desirable to gain information about the nature of the underlying phenomena. It is important to search for new interfacial phenomena and their elucidation, development of new techniques, instrumentation and procedures, improving sensitivity and selectivity of measurements, and widening scope of applicability of the method.

In parallel with the development of adsorption voltammetry goes the development of electrocapillary adsorption measurements [1]. The sensitivity and reproducibility have recently been improved, mainly due to the method of drop-time measurement with controlled convective accumulation [11-14, 47-49, 52], i.e. a method based on principles similar to AdSV. Direct analytical applications of this method have currently been tested.

The present problems include new ways of interfacial accumulation of surface-active substances [19, 41, 49, 53], possibilities of the determination of the polarity of surfactants [14], measurements of the total contents and activities of surface active substances in waters and solutions [14, 33], modification of electrosorption activity of substances by the action of ultrasound [49, 54-56], phenomena connected with the use of renewed miniaturized electrode systems [14], the possibility of "absolute" measurements based on the high reproducibility of the electrode size [12, 35, 43-49], effects noxious to sensors during vibrations [12, 14, 41, 54, 55], at extreme negative potentials [12, 14, 49], under the influence of strong external fields of energy [63], etc.

A complex solution to such problems represents the basis of a new era of development of electrosorption methods.

3.2 Effects of interfacial activity in adsorption voltammetry and related methods and their utilization in analytical applications

3.2.1 Experimental arrangements

A substance dissolved in a solution displays an interfacial activity at an electrode when it adsorbs at the electrode surface, i.e., when its relative surface excess (sometimes not quite correctly called "surface concentration" - see Eq.(39)), Γ_i at the electrode/solution interface is higher than its bulk concentration in the solution [64, 65, 14, 18]. The adsorption may be either non-specific, e.g., adsorption of solvated ions in the outer Helmholtz plane (2 in ***Fig.1***) at the electrode, or specific, in the inner Helmholtz plane (1 in ***Fig. 1***), where the short-range forces come into play accompanied by a partial overlap of the electronic orbitals with the electrode. In the latter case, the electrode becomes, in principle, chemically modified.

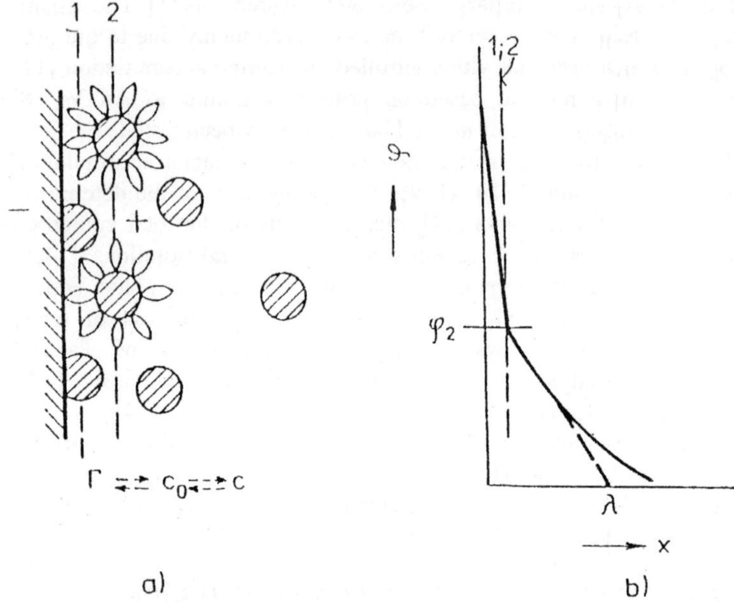

Fig. 1: **a) Electrode/solution interface with non-specifically adsorbed (solvated) particles in plane 2 and with specifically adsorbed particles in plane 1; b) Example of potential gradient φ in the close proximity of the electrode**

λ - width of the diffuse double layer with potential φ_2

In electroanalysis utilizing adsorption-desorption effects, the tendency is always to provide conditions under which the studied substance exhibits the highest specific adsorption while the other solution components are surface inactive. If the concentration and the surface activity of the solution background are high, the surface excess Γ of the studied component is low. It has been established experimentally that measurable changes of the surface activity expressed, e.g., in terms of Γ or interfacial tension Γ (cf. Eq. (3), (6) *et seq.*) or a corresponding electrical signal (cf. Eq. (40) *et seq.*), can only be caused by those particles or their functional groups which, at least under certain conditions, participate markedly in the formation of the double layer structure [3, 64, 65, 14]. For these reasons, e.g., the surface activity of micellar solutions [14, 66, 67], where the micelle size exceeds the dimensions of the double layer, barely increases with increasing concentration of micelles. On the other hand, strong surface activity can be observed in solutions of large biologically active molecules [68, 69, 24] which are attached to the electrode surface through an interaction of some functional groups, heteroatoms, etc.

Important for analytical applications are, e.g., the recording of the current response to potential-induced adsorption or desorption of a given compound as a function of its concentration, or the recording of the concentration dependence of the surface tension (or, better interfacial tension) of mercury or of its functions (following the changes in the differential capacity of the electrode, etc.).

A suitable experimental arrangement for the curve recording in AdSV is shown schematically in ***Fig. 2***. It consists of a polarograph (1) permitting d.c. polarization, or, better, differential pulse polarography (DPP), "fast scan DPP", or differential pulse voltammetry (DPV)[1]); a renewed stationary mercury electrode (2) and a cell (3) with the solution with a reference electrode (4), sometimes also an auxiliary electrode (5), a stirrer (6) and an inert gas inlet (7).

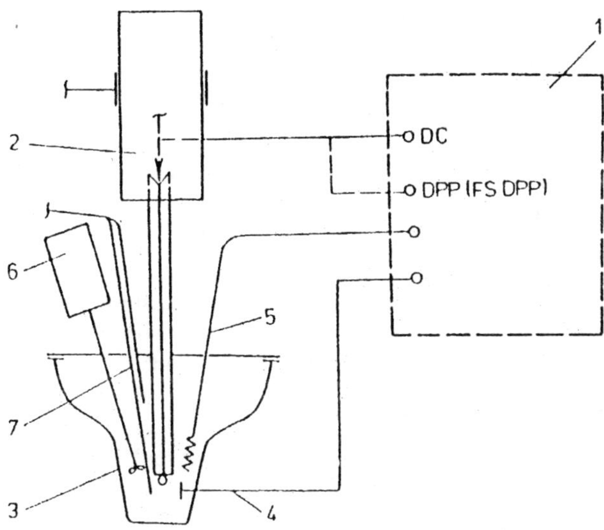

Fig. 2: Scheme of experimental set-up for adsorption voltammetric measurements (description in the text)

[1]) According to the IUPAC nomenclature, the term "polarography" is reserved for methods using electrodes with the surface area changing with time, whereas the methods using electrodes with time-invariable surface areas are termed "voltammetry".

A more up-to-date, computer controlled arrangement, either in a bench or in a portable version [44-49, 70-73], can be represented by the first Czech commercially available Eco-Tribo Polarograph PC-ETP. The design features, apart from the appropriate software, new types of miniaturized electrode systems (a pen-type mercury electrode and others), new ways of evaluation of the experimental data, special methods and accessories for voltammetric analysis in tribology, and miniaturized modular components. The polarograph is designed as a three-electrode instrument, however, due to the miniature size of the electrodes the currents are so small that the use of two electrodes, i.e. of the working and the reference electrode, is completely satisfactory in measurements in aqueous media containing a supporting electrolyte.

The old drop-time method has been adapted for a sensitive direct measurement of the surface (interfacial) tension of mercury [11-14], where the surface tension is again determined from the drop-time of spontaneously falling mercury drops. The difference from the classical method is the use of the spindle capillary [12, 35, 47-49, 18, 74], of the controlled convective adsorption accumulation achieved by precisely timed stirring during the growth of the drop [11-14], and of new programmed automated regimes, synchronized with the moment of detachment of each preceding drop. By this modification, the drop-time utilizable for measurement has been prolonged more than ten times, the reproducibility improved 100 times with a simultaneous improvement in the stability and sensitivity (the detection limit decreases from 10^{-5} to 10^{-7}- 10^{-8} mol l^{-1}).

A scheme of an example [14] of experimental arrangement is shown in ***Fig. 3***. The set-up consists of the polarization circuit (A) containing a source of the polarizing voltage (1), variable resistor (2), measuring electrode (3) and reference electrode (4). In parallel are connected the detection circuit (B) containing an a.c. generator (5) of a low-amplitude (e.g. 5 mV) and high-frequency (e.g. 20 x 10^3 s^{-1}) voltage, a control circuit (6) for the stirrer (7), adjustable resistor (8), measuring electrode (3), electric stop-watch (9), auxiliary electrode (10) and a capacitance filter (11). The filter prevents passage of the d. c. signal beyond polarization circuit A. The values of the resistances 2 and 8 are chosen so that the a.c. signal does not practically pass through A and its level at the electrode/solution interface remains minimal. The drop-time is then detected by recording the discontinuous change in the electrical resistance due to the discontinuous decrease in the electrode area at the moment of drop detachment. The proper measurements by the drop-time method with controlled adsorptive accumulation at very low surfactant concentrations are carried out in stirred solutions, e.g., for about 90 % of the

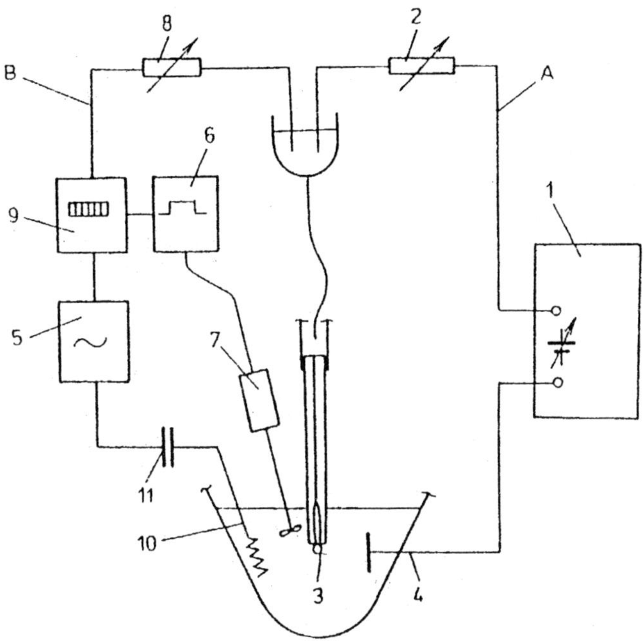

Fig. 3: **Experimental arrangement for electrocapillary measurements by the drop-time method with convective adsorption accumulation (description in the text)**

drop-time t. The aim is then to determine the dependence of the drop-time on the applied voltage E, i.e. $t = t(E)$ at a constant concentration c, or the dependence $t = t(c)$ at a constant E. The mercury surface tension Γ can be easily calculated (see below) from time t, which is the main step in quantitative evaluation of the adsorption measurements - the determination of adsorption isotherms, adsorption parameters of the surfactants, etc.

3.2.2 Reversible and quasi-reversible adsorption, its characteristics and manifestations

In a reversible adsorption process the adsorption-desorption step is very fast compared with other preceding or follow-up events (diffusion, potential change, change of the electrode area, etc.), and it is the rate-determining factor in the adsorption mechanism. The transition between reversible and irreversible adsorption and chemisorption is not sharp and the determining character of the process depends on the experimental conditions. This follows from the relation between the time τ during which the molecule remains in the adsorbed state at temperature T, and the adsorption energy ΔG [66]

$$\tau = \tau_0 \exp(\Delta G/RT) \qquad (1)$$

(under usual conditions, $\tau \approx 10^{-12}$ s). For physical adsorption, the values of τ are between 10^{-7} and 10^{-12} s, whereas in chemisorption they can be hundreds and more seconds. The order of these values has to be kept in mind when considering experimental conditions in AdSV.

3.2.2.1 ELECTROCAPILLARITY AND ADSORPTION VOLTAMMETRY

The dependence of the mercury surface tension $(\Gamma \sim t)$ in solution on the applied potential E at a given solution composition is known as the electrocapillary curve. For dilute solutions (up to concentrations of cca 0.2 mol l^{-1}) of surface inactive supporting electrolytes, the following relation is approximately valid (for more rigorous relations see, e.g., [65]):

$$\Gamma \approx \Gamma_z - k' \sqrt{c}\, (E - E_z)^2 \qquad [T = \text{const}] \qquad (2)$$

where Γ_z and E_z (thereby $\Gamma_z \equiv \Gamma_{E_z}$) are the values of Γ and E at the electrocapillary maximum, c - concentration of the electrolyte, T - temperature and k' - a constant. The approximately parabolic shape of the electrocapillary curve of the supporting electrolyte is shown in ***Fig. 4a*** (curve 1). Gradual additions of a surfactant (curves 2 and 3) result in a deformation of the original shape of the Γ - E dependence [64, 65, 18, 21]. For nonpolar neutral compounds, the position of the maximum E_z remains thereby approximately constant, for cation active compounds it shifts to positive and for anion active ones to negative potentials. The potentials E_a and E_d , at which the electrocapillary curves before and after addition of the surfactant merge, correspond to the positions of the voltammetric desorption and adsorption peaks (curves 4, 4', 5). A point of special significance is the potential of maximum adsorption E_{max} where the interfacial activity of the given surfactant in the given medium reaches its highest value. The corresponding values of the charge density on the electrode surface, $q = -(\partial \Gamma / \partial E)_c$, the composition and differential capacities of the electric double layer C as functions of the potential are shown in ***Figs. 4b*** and ***4c*** [14, 16].

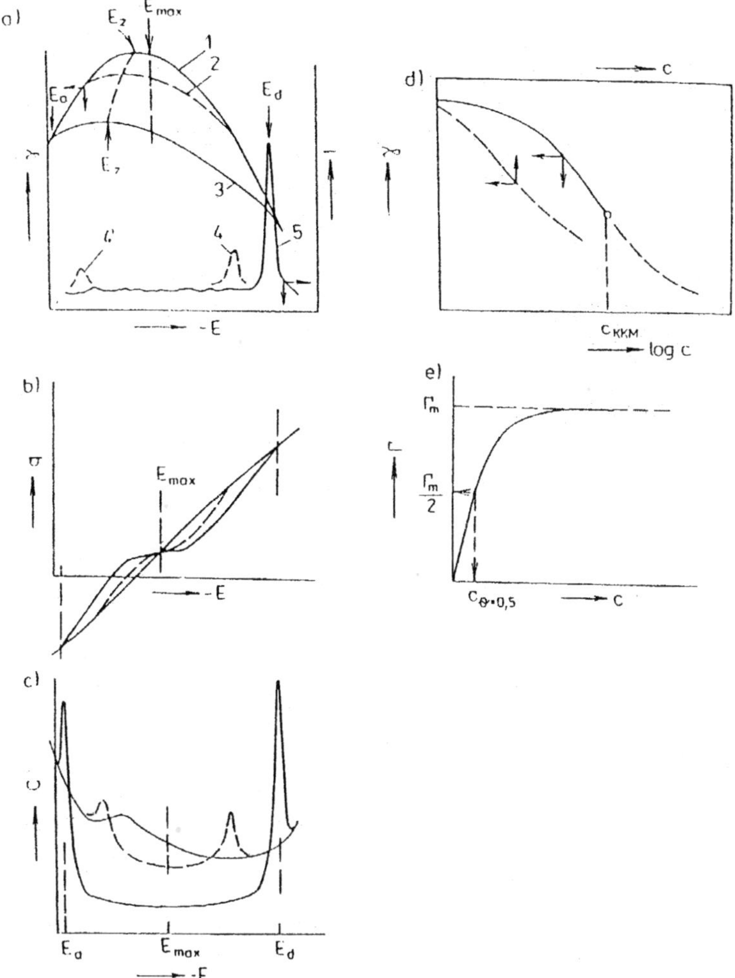

Fig. 4: **a) Electrocapillary curves *Γ-E* and *i-E* curves as recorded in the method of adsorption voltammetry (AdSV)**
b, c) Potential dependences of charge density *q* and differential capacity *C*
d) Concentration dependence of interfacial tension of mercury *Γ*;
c_{ccm} - critical concentration of formation of micelles, precipitates etc.
e) Adsorption isotherm - dependence of surface concentration *Γ* on bulk concentration *c*

The time changes of charge density q at an electrode under a constant composition of the solution are the actual cause of the corresponding current response measured in AdSV (cf. ***Fig. 4a***, curves 4, 4', 5). Directly utilizable concentration dependences involve the voltammetric and capacitance peak heights and Γ, t or $\Delta t/t = (t_0 - t)/t_0$ in the region $E \approx E_{max}$ (t_0, t stand for the drop time before and after the addition of surfactant of concentration c). The equilibrium values of Γ and t are mutually directly proportional [14, 60, 75]:

$$\Gamma = \frac{\gamma_{ref}}{t_{ref}} \frac{\rho_{Hg} - \rho}{\rho_{Hg} - \rho_{ref}} Kt \tag{3}$$

where Γ_{ref}, t_{ref} and density ρ_{ref} are referred to a reference solution, ρ_{Hg} is the density of mercury, ρ the density of the measured solution and K is a constant ($K \approx 1$). Dilute solutions of "inactive" supporting electrolytes usually serve as reference solutions (0.1 mol dm^{-3} $NaHSO_4$, K_2SO_4, KOH, NaOH, KF, NaCl, etc.) for which for $(\Gamma_{ref})_z$ [at $q = 0$, $(\Gamma_{ref})_z \equiv (\Gamma_{ref})_{Ez}$] a value is substituted which can be calculated from the relation (for Γ in mN m^{-1} and t' in $^{\circ}$C) [64, 14]

$$(\Gamma_{ref})_z = 426.7 - 0.17(t' - 18) \tag{4}$$

For waters or supporting electrolytes polluted by surfactants, the following approximate equation holds [17, 18, 24, 14]

$$\Gamma \approx [(\Gamma_{ref})_z / \Gamma(t_{ref})_z]t \tag{5}$$

where the value of $(\Gamma_{ref})_z$ is obtained from Eq. (4). The dependence of $\Delta t/t$, t or Γ on log c at constant E is usually hyperbolic (cf. ***Fig. 4d***) and depends little on the potential around E_{max}. We speak about congruence of the curves with respect to potential [17, 77]; its analytical consequence is a small potential dependence of the sensitivity and the detection limit of the described methods. Inflexion points, lags or breaks on the t, Γ or $\Delta t/t$ *vs.* log c curves indicate, as a rule, qualitative changes in the system. The position of an inflexion point, e.g., at $c = c_{ccm}$ (***Fig. 4d***), often corresponds to the beginning of micelle formation [24, 66]. The adsorption is then no longer considered as reversible or at equilibrium, determined solely by one electrical (E or q) and one concentration (c or Γ) variable [12, 14].

From the $\Gamma \sim \log c$ dependence it is easy to determine the corresponding adsorption isotherm (***Fig. 4e***) and the characteristic adsorption parameters [12, 17, 18] for the given substance using Eq. (13), given medium and other experimental conditions (potential, temperature, etc).

3.2.2.2 FUNDAMENTAL RELATIONS AND THEIR THERMODYNAMIC BACKGROUND

On the basis of direct electrocapillary studies of strongly surface active substances (SAS) using the described method [11-14, 17, 15, 52] over a wide range of concentration and an analysis of the data obtained [12], it was demonstrated that the electrocapillary diagram $\Gamma = \Gamma(E,c)$, i.e., a series of $\Gamma = \Gamma(E)$ dependences for different surfactant concentrations c in a given supporting electrolyte, thermodynamicaly characterizes the studied equilibrium system. It has the character of a state diagram [12, 14, 78] and offers information about the possibilities of analytical utilization of interfacial measurements [14, 78]. The behaviour of the system is quantitatively described by the Gibbs-Lippmann equation [64, 65, 17] expressing the state before addition of the adsorptive substance of concentration c (under given temperature and pressure) [Eq. (6)] and after addition [Eq. (7)]:

$$\mathrm{d}\Gamma_0 = -q_0\,\mathrm{d}E - \sum_j \Gamma_j\,\mathrm{d}\mu_j \qquad (6)$$

$$\mathrm{d}\Gamma = -q\,\mathrm{d}E - \Gamma\mathrm{d}\mu - \sum_j \Gamma_j\,\mathrm{d}\mu_j \qquad (7)$$

where Γ_j are the relative surface excesses and μ_j the chemical potentials of the supporting electrolyte components.

The physical meaning of variable Γ_j follows [64-66, 60] from the relation

$$\Gamma_j = \Gamma_j^* - \Gamma_{H_2O}^* (x_j/x_{H_2O}) \qquad (8)$$

where

$$\Gamma_j^* = n_j/A; \qquad \Gamma_{H_2O}^* = n_{H_2O}^*/A \qquad (9)$$

Γ_j^* and $\Gamma_{H_2O}^*$ represent number of mols of the j-th solution component or of water per unit of electrode area, respectively, Γ_j stands for surface excess of the j-th component at the electrode/solution interface compared with its relation to the amount of water in the bulk of the solution; x_j, x_{H_2O} are the mole fractions of the above components in the bulk. In the case of aqueous solutions of uni-univalent supporting electrolytes, Γ may be taken for each ion separately or for relative surface excess of the salt (BA) Γ_S [65, 60] of mean activity $a_\pm$. For $q = 0$ and hence for $E = E_z$, relation (10), analogous to Eq. (6), is obtained

$$\Gamma_S = \Gamma_{B+} + \Gamma_{A-} = -1/2RT\,(\partial\Gamma/\partial\ln a_\pm)_{E_z} \qquad (10)$$

The value of $(\partial\Gamma/\partial\ln a_\pm)$ is an important criterion for selecting appropriate inactive supporting electrolytes.

For the so-called surface pressure [77, 11, 65], $-d\pi = \Gamma_0 - \Gamma$, it follows from Eqs (6) and (7)

$$-d\pi = d\Gamma - d\Gamma_0 = -(\Gamma - \Gamma_0)\, dE - \Gamma d\mu \qquad (11)$$

where $d\mu = 2.303\, RT \log c$ (for $c \rightarrow 0$), and

$$\pi = \pi(\beta, \Gamma_m, c, E, a_i, \ldots) \qquad (12)$$

Relation (12) expresses the general dependence of measurable changes of interfacial activity on the concentration, potential, basic adsorption parameters - adsorption coefficient β and highest surface concentration Γ_m, and on other factors like Frumkin's interaction coefficient, solvation constants, etc. A number of theoretical and practical consequences can be derived from the above equations: The value of Γ can, e.g., be determined for constant E and a given supporting electrolyte composition (μ_j = const, $d\mu_j = 0$)

$$\Gamma = \frac{1}{2.303RT} (\partial \Pi / \partial \ln c) \qquad (13)$$

where $-d\Gamma = d\pi$. Further it follows that

$$(\partial q / \partial \Gamma)_E = RT (\partial \ln c / \partial E)_\Gamma \qquad (14)$$

$$(\partial \Pi / \partial E)_c = (1/RT)\, (\partial q / \partial \ln c)_E \qquad (15)$$

The informative contents of these and the following relations will become clear below. For charge density q and differential capacity C we obtain

$$q = -(\partial \Gamma / \partial E)_{\mu,\mu_j};\ C = (\partial q / \partial E)_{\mu,\mu_j} = (\partial^2 q / \partial E^2)_{\mu,\mu_j} \qquad (16)$$

The assumption [12, 14] that the equilibrium electrocapillary diagram $\Gamma = \Gamma(E,c)$, shown schematically in ***Fig. 5***, possesses the properties of a diagram of state (analogous to the well known p-V-T diagram for gases) below the curve of the supporting electrolyte allows us to derive some practically useful relations and coefficients [14, 78]. We usually start by expressing the changes in one selected variable in terms of changes in two other variables, one expressing the concentration and the other the electrical properties, as, e.g.

$$\gamma = \gamma(E,c); \qquad \Gamma = \Gamma(E,c); \qquad q = q(E,c); \qquad \Gamma = \Gamma(q,\Gamma) \qquad (16a)$$

etc.. Hence,

$$\partial \Gamma = (\partial \Gamma / \partial E)_{\ln c}\, dE + (\partial \Gamma / \partial \ln c)_E\, d \ln c = -q\, dE - \Gamma d\mu \qquad (17)$$

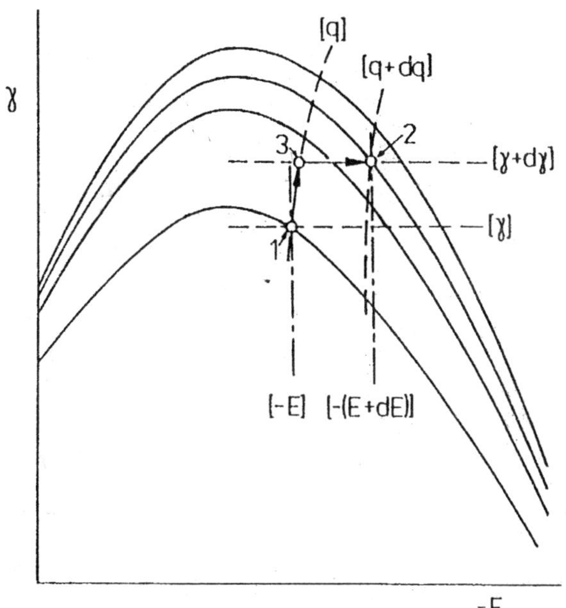

Fig. 5: **Transition of an electrocapillary system from state 1 to state 2 along the path 1 → 3 [*q*] and 3 → 2 [*Γ*]. The change of state from point 1 to point 2 is independent of the path**

and so on. From the assumption that the *Γ-E-c* diagram has the character of a thermodynamic diagram of state it follows, that the value of Γ at a given point of the diagram may be determined from the changes at constant Γ, E, or q, as

$$\Gamma_\Gamma = -q(\partial E/\partial\mu)_\Gamma; \qquad \Gamma_E = -(\partial\Gamma/\partial\mu)_E; \quad \Gamma_q = -[\partial(\Gamma + qE)/\partial\mu]_q \quad (18)$$

while for a reversible system

$$\Gamma_\Gamma = \Gamma_E = \Gamma_q \quad (19)$$

The extent of quasi-reversibility (more precisely of irreversibility) of the system can be expressed quantitatively in terms of the deviations, e.g., from a straight line with slope of 45° in the diagram Γ_q - Γ_E [12, 78].

For capacity C [77, 14] it holds that

$$q = q(E,\Gamma); \quad dq = (\partial q/\partial E)_\Gamma dE + (\partial q/\partial\Gamma)_E\, d\Gamma \quad (20)$$

$$C = \mathrm{d}q/\mathrm{d}E = (\partial q/\partial E)_{\Gamma} + (\partial q/\partial \Gamma)_{E}\,(\mathrm{d}\Gamma/\mathrm{d}E) = C_{\mathrm{p}} + C_{\mathrm{dop}} \qquad (21)$$

where the component C_p at Γ = const. is known in the literature as the true capacity and the component C_{dop} as the complementary capacity. In course of sudden changes of surface concentration Γ the component C_{dop} markedly increases which gives rise to analytically utilizable peaks [16, 77] on *C-E* curves.

From the assumption of equality of the chemical potentials of substances in adsorbed state and in the solution there follows the basic condition for the adsorption isotherm [77]:

$$\beta c = f(\theta) \qquad [E = \mathrm{const}] \qquad (22)$$

with

$$\Delta\overline{G} = RT \ln \beta; \qquad \theta = \Gamma/\Gamma_{\mathrm{m}} \qquad (23)$$

where θ is the degree of coverage of the electrode and $\Delta\overline{G}$ - the free energy of adsorption. For practical correlations of the experimental results, the Langmuir (Eq. (24)) and Frumkin's isotherms (Eq. (25)) have been used most often [64]

$$\frac{\theta}{1-\theta} = \beta c \qquad (24)$$

$$\frac{\theta}{1-\theta}\exp(-2a\theta) = \beta c \qquad (25)$$

where coefficient a is known as Frumkin's coefficient characterizing the attraction or repulsion of the adsorbed particles. In the absence of substitution and other interaction phenomena (see below), the value of a is zero or is very small. A combination of the differential forms of Eqs. (17) and (22) with the model of charge contribution additivity leads [76] to equations expressing additivity of contributions of charge densities on an unoccupied (q_0, at θ = 0) and a fully occupied (q_1 , at θ = 1) electrode surface:

$$q = RT\Gamma_{\mathrm{m}}\,(\mathrm{d}\ln\beta/\mathrm{d}E)\,\theta + q_0 \qquad (26)$$

$$q = q_0(1-\theta) + q_1\theta; \qquad \theta = \Gamma/\Gamma_{\mathrm{m}} \qquad [E = \mathrm{const}] \qquad (27)$$

An exact derivation of relations (26) and (27) assumes that the adsorption coefficient β depends only on potential, i.e., that it does not change with the degree of coverage θ, which is not always fulfilled, especially at higher concentrations. The potential dependence $\beta = \beta(E)$ in the neighbourhood of $E = E_{max}$ can be expressed approximately [64, 65,18]

$$\beta = \beta_{max} \exp[\alpha(E - E_{max})^2] \tag{28}$$

where α = const. The potential dependence of β can be rapidly determined from a series of Γ-log c dependences for different E using Eq. (22), on taking its logarithm for θ = const [17]:

$$\log \beta = -\log c + \log[f(\theta)] = -\log c + \text{const} \tag{29}$$

In real systems, substitutions of other adsorbed components from the medium (including components of solvents, supporting electrolytes, etc.) by particles of the SAS, interactions of SAS with these components and other kinds of further interparticle interactions and processes have to be considered in addition to the adsorption-desorption processes proper [17, 64]. Therefore a general form of adsorption isotherm was used for description of SAS adsorption [78, 79, 80], e.g.,

$$f_1(\theta) \exp[f_2(\theta)] = \beta(c - c_0)\ . \tag{30}$$

and, in particular,

$$\frac{\theta}{1-\theta} \exp\left(\sum_{j=1}^{n} A_i \theta^i\right) = \beta(c - c_0) \tag{31}$$

i.e., in special cases of Eqs (32) and (33), where $f_1(\theta)$ and $f_2(\theta)$ stand for general functions of θ; A_i and c_0 are coefficients and n is the degree of the polynomial. The expressions (30) and (31) include relations (24) and (25) and are in agreement with Eqs (22) and (23).

The following semiempirical model, Eqs (32) - (36), can be used for fitting, characterization or description of the discussed systems at least within a limited range of variables [78, 79]. The x and y variables may have, according to the actual need, various meaning, as shown further.

$$f_3(Y) \exp[f_4(Y)] = x - x_0 \tag{32}$$

in a particular example

$$\frac{Y}{1-Y} \exp\left(\sum_{j=1}^{n} A_i Y^i\right) = x - x_0 \tag{33}$$

or

$$y - y_0 = f_5(X) \exp[f_6(X)] \tag{34}$$

in a particular example

$$y - y_0 = \frac{X}{1-X} \exp\left(\sum_{j=1}^{n} A_i X^i\right) \tag{35}$$

where meaning of the symbols is the following

$$Y = (y - y_0)/(y_m - y_0); \qquad X = (x - x_0)/(x_m - x_0) \tag{36}$$

y - independent variable; x - dependent variable; x_0, y_0 - coordinates of the origin of the y *vs* x dependence; x_m, y_m - the highest values of the x, y variables; A_i - constants (parameters) of the y *vs* x dependence; $f_3(Y)$, $f_4(Y)$, $f_5(X)$, $f_6(X)$ - general functions of Y and X. The variables y or x may have the physical meaning of electric current i, potential E, time t, concentration c, relative surface excess Γ or coverage θ, interfacial tension Γ, electric capacity C, time of accumulation t_{acc} (or t), potential of accumulation E_{acc}, etc., of their differences, derivatives or integrals, the absolute values of their differences, or of other functions. The fitting of curves like i-E, i-t, i-c, i-t_{acc}, etc., with parameters which may have physical meaning [14] can serve as an example. Their shapes are often similar to voltammetric peaks, or they are parabolic, hyperbolic or S-shaped (potentiometric [20]) curves.

One of the further consequences of the thermodynamic character of electrocapillary systems is the expression and analysis of the properties of the E - log c dependences for varying q, or the $(\partial E/\partial \mu)_q$ coefficient, especially for very dilute surfactant solutions. On the basis of the above concept, the coefficient has been expressed [12, 78] as

$$(\partial E/\partial \mu)_q = (\partial q/\partial \Gamma)_E \Gamma / C_\Gamma - \left[\frac{q(\partial q/\partial \gamma)_E}{C_\gamma}\right] \tag{37}$$

which, for $q = 0$, $E = E_z$, when,

$$\left[\frac{q(\partial q/\partial \gamma)_E}{C_\gamma}\right] = 0; \quad C_\Gamma = C$$

assumes the form

$$(\partial E/\partial \mu)_q = \left[\frac{(\partial q/\partial \gamma)_{E_z}}{C}\right]\Gamma = -\left[\frac{(\partial q/\partial \mu)_{E_z}}{C}\right] \tag{38}$$

The application of this new electrocapillary method has not confirmed the previous assumptions [81, 82] about constancy of the $(\partial E/\partial \mu)_{q=0}$ coefficient for surface active substances (SAS) and has shown [12, 78] that the dependence E - log c at $q = 0$ is S-shaped for a wide concentration range of SAS which can be approximated by a straight line with the slope $(\partial E/\partial \log c)_{q=0}$ over a narrow range. The validity of relations (37) and (38) has been confirmed. The results are applicable, e.g., for the determination of polarity of SAS.

3.2.2.3 SIGNIFICANCE OF THE ABOVE RELATIONS

The properties of the described adsorption system are near to those of aqueous electrolytes containing traces of strongly adsorbable substances at concentrations corresponding to the degree of electrode coverage between about $\theta = 0.1$ and 0.7. From the practical point of view [18, 83], substances with adsorption coefficients higher than $\beta \geq 1 \times 10^4$ - 5×10^4 l mol^{-1} are considered as strongly adsorbable. The value of β can readily be estimated as $\beta \approx 1/c_{\theta=0.5}$ from the adsorption isotherm (cf. ***Fig. 4e***). Strongly adsorbable surfactants appear especially prospective for application of adsorption voltammetry and interfacial measurements. This has been confirmed in voltammetric determinations of traces of saponates in aqueous solutions [7, 11, 104], trace concentrations of oil products [21, 22, 25], diazepam [26], bipyridyl and its derivatives (herbicides) [28, 29], carcinogenic cyanuric chlorides and tris-biphenyls [84], low concentrations of metals bound in adsorptive complexes [8,105,109,110], capacitive determination (by means of *C-E* curves) of polyoxyethylenealcohols and a series of surfactants [16], electrocapillary determination (by the *t-c* method) of starch in waters [24], etc.

By determining the polarity of coefficient $(\partial E/\partial \log c)_{q=0}$ for $c \rightarrow 0$, the polarity of a surfactant or of a dominating component in their mixture can be estimated.

An important criterion of surface activity of supporting electrolytes is the value and the sign of the coefficient $(\partial \Gamma/\partial \ln a_{\pm})_{E_z}$ in this medium at the potential of electrocapillary maximum where the net surface charge of the electrode equals zero. In the past [1] it was assumed that fluorides, sulphates, chlorides, nitrates, hydroxides and other salts of alkali metals are almost or completely surface inactive in a potential region near $E = E_z$, with virtually the same value of $(\Gamma_{ref})_z$ (cf. Eq. (4)). This idea was especially strongly held for the solutions of alkali fluorides. It was demonstrated for the first time [60], using the new method of electrocapillary measurements complemented by the use of polyethylene capillaries [57-61] resistant to fluoride corrosion, that fluoride solutions exhibit a relative surface excess $\Gamma_S < 0$, especially at higher concentrations, because their surface concentration Γ_S^* is then less than the term $\Gamma_{H_2O}^*$ (x/x_{H_2O}). In a similar way it was possible to determine a sequence of supporting electrolytes according to their specific surface activity for lower concentrations at $E = E_z$, as shown in [85].

In development of electrosorption methods for determination of trace concentrations of surfactants and for theoretical research, it is necessary to consider possible specific adsorption of supporting electrolytes.

For SAS, the relative surface excess of which is denoted as Γ in Eq. (7), the form of relations (8) and (9) is considerably simplified, as in the expression,

$$\Gamma = \Gamma^* - \Gamma_{H_2O}^*(x/x_{H_2O}) \quad (39)$$

where the mole fraction of SAS in the bulk $x << x_{H_2O}$ and hence their relative surface excess Γ is practically identical with the surface concentration Γ^*. Therefore, Eq. (19) can be used as a quantitative criterion of reversibility of adsorption [12, 78].

Height, shape and potential shift of the voltammetric peak

The voltammetric signal in AdSV has some specific properties. Unlike the faradaic reactions, adsorption processes produce changes in the course of the curve of the supporting electrolyte in the potential region of the peak, noticeable even at lowest surfactant concentrations. However, with increasing value of the signal these changes play progressively a less important role.

The recorded current signal can be expressed in terms of a relation [86, 14] which is a voltammetric analogy of Eq. (21)

$$I = dq/dt = C_p(dE/dt) + (\partial q/\partial \Gamma)_E(d\Gamma/dt) \quad (40)$$

Its magnitude thus markedly depends on the time change of the surface concentration Γ, and hence also on the adsorption energy and its changes. The current increases with the rate of potential change, dE/dt, and with the value of adsorption energy ΔG. The properties of the coefficient $(\partial q/\partial \Gamma)_E$ follow from Eqs (26) and (27).

For a current related to the supporting electrolyte curve it holds [14] that

$$I \sim [(\partial q/\partial E) - (\partial q_0/\partial E)]_{\mu,\mu_j} = \left[\frac{\partial(\partial\pi/\partial E)}{\partial E}\right]_{\mu,\mu_j} \quad (41)$$

$$I \sim (C_1 - C_0)\theta + (q_1 - q_0)(d\theta/dE) \quad (42)$$

where the indexes 1 and 0 refer to the states $\theta = 1$ and $\theta = 0$, respectively.

A frequently used way of determining the height of a peak, especially in the absence of the supporting electrolyte curve recorded under the same conditions, is to measure the distance from the straight line connecting the two intercepts of the peak with the baseline. A comparison of this manner of evaluating the signal with the usual way of measuring with respect to the supporting electrolyte is demonstrated in ***Fig. 6*** for analysis of motor oil by the AdSV - "fast scan" differential pulse voltammetry (FS DPV method) (cf. [22]). It is obvious that the measurement with respect to the supporting electrolyte produces lower detection limits.

The measured signal has sometimes a sharp needle-like shape. Such effects, observed, e.g., in voltammetric determination of oil fractions in water [22, 25], can be explained by sudden structural changes of the double layer or of adsorbed films, by reorientation or by phase changes in the adsorbed layer.

The signal sometimes acquires an even more complex shape, e.g. a double peak [22, 25, 30, 104]. This phenomenon is obviously connected with a concentration dependent adsorption and/or desorption mechanism coming into play in the course of the potential change (due to reorientations, different activity of individual parts of the molecule, etc.); as a rule it depends on the temperature and experimental conditions.

An AdSV voltammetric curve can be rather complex when the electrode process involves not only adsorption and charge transfer, but also factors like interactions in the double layer, dimerization or other chemical reactions, ion pair formation, etc. In analytical applications of AdSV it is, therefore, essential to find out and maintain appropriate simple experimental conditions. For example AdSV of 4,4'-bipyridyl [28] yields complicated curve shapes in acidic solution whereas in alkaline media a single well-developed peak occurs at -1.3 V (*vs.* SCE) which is suitable for analytical purposes.

One of the characteristic features of peaks connected with desorption is the experimentally found shift of their potential E_p (or ΔE_p) [8, 22, 25], corresponding to the shift of the desorption potential E_d (or ΔE_d) (***Fig. 4a***), with increasing concentration of the surfactants. It is most marked at low concentrations. On the basis of the assumption about the state behaviour of the studied adsorption system (see part 3.2.2.2, ***Fig. 5***) [12, 14, 78] the relation (43) was derived for the shift ΔE_p in dependence on the concentration c

$$(\Delta E_p)^2 = k_1 \log c + k_2 \qquad (43)$$

In principle, from Eq. (43) follows relations (43a)-(43d):

$$d\Gamma = -q\, dE_d - \Gamma d\mu; \qquad \Delta E_d \equiv \Delta E_p \qquad (43a)$$

$$(\partial \Delta E_d / \partial \mu)_q = (\partial \Gamma / \partial q)_\mu = \Delta\Gamma/\Delta q = \Delta\Gamma / C\Delta E \qquad (43b)$$

$$\Delta E_d\, d\, \Delta E_d = \tfrac{1}{2} d(\Delta E_d)^2 = (\Delta\Gamma/C) \times 2.303\, RT \log c \qquad (43c)$$

$$(\Delta E_d)^2 = (\Delta E_p)^2 = k_1 \log c + k_2 \qquad (43d)$$

where k_1 and k_2 are constants.

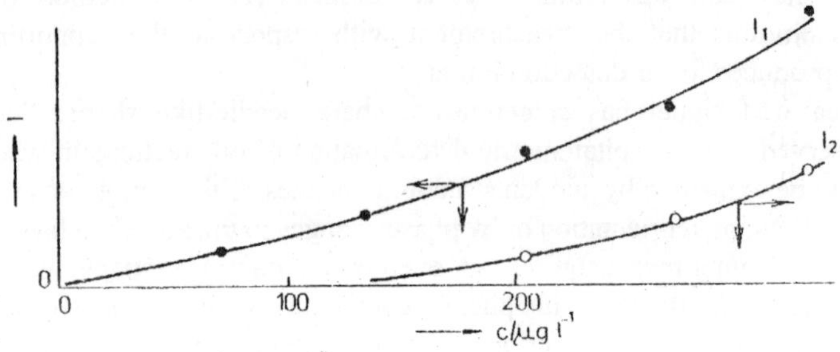

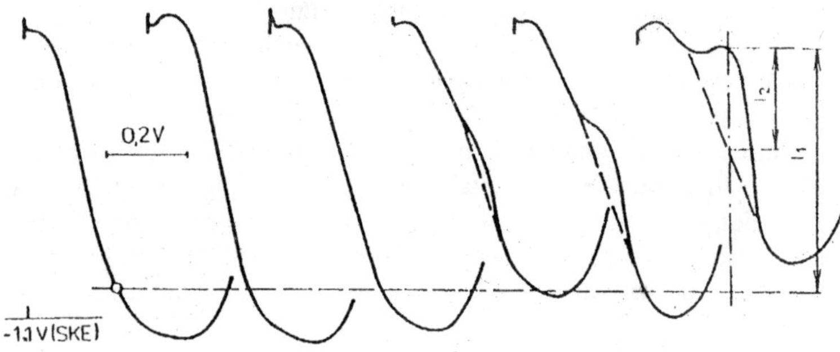

Fig. 6: **Two ways of evaluation of the voltammetric signal due to traces of motor oil in 0.1M KCl**

I_1 - measured from the curve of supporting electrolyte, I_2 - measured from the line connecting the two intercepts of the peak. According to [22]

Sensitivity, detection limit and relative adsorptivity

The properties of these analytical parameters can best be interpreted [14, 78] in the Henry concentration region (for low values of c), where the isotherm (24) is approximately linear:

$$\theta = \Gamma/\Gamma_m = \beta c; \qquad d\Gamma = RT\,\Gamma_m\, dc \tag{44}$$

It follows from Eq. (5) that $d\Gamma/\Gamma = dt/t$ and thus for small Δt it holds that

$$x_x = \Delta t/t = (RT/\Gamma)\Gamma_m \beta c; \qquad c \to 0 \qquad [E = \text{const}] \tag{45}$$

The sensitivity of electrocapillary measurements for lowest concentrations is thus directly proportional to the adsorption coefficient. Its maximum value is reached for $E = E_{max}$, when, according to Eq. (28), $\beta = \beta_{max}$. Moreover, the value of the coefficient $(\partial\Delta t/\partial\Delta c)_E$ increases with increasing drop-time t for E = const.

For strongly surface-active substances in various solutions of inactive electrolytes, the sensitivity of equilibrium electrocapillary measurements for $c \to 0$ is almost independent of the kind of electrolyte and hence is, in aqueous solutions under given thermodynamic conditions (temperature, pressure, etc), characteristic for the given substance. As the differences in the values of Γ_m for different surfactants are considerably less than the differences in the values of the adsorption coefficients β, a comparison of limiting sensitivities $(x/c)_{E,c\to 0}$ provides a rapid orientation about the relative interfacial activity (adsorptivity) of a substance in comparison with a selected reference substance $(\beta'/\beta'_{ref}) = \beta\Gamma_m/(\beta\Gamma_m)_{ref}$. With a known value of the coefficient $[(\Delta t/t)/c]_E$ for a reference system, only one measurement is sufficient for determining (β'/β'_{ref}), selected for such a surfactant concentration c, where the relative changes $\Delta t/t$ fall between 0.01 and 0.05. The result of the comparison is independent of the parameters of the apparatus, i.e., the drop-time, etc. The selected potential E is usually equal to E_{max} or E_z of the supporting electrolyte. If the detection limit c_{min} is substituted for c in Eq. (30), it is found that its value is lower for higher values of β and Γ_m coefficients and for a longer drop-time.

The potential-dependence of the limiting sensitivity and detection limit is obvious from Eqs (45) and (28). For the detection limit it can also be written (for $\pi = \pi_{min}$) that

$$\left(\frac{\partial \ln c_{min}}{\partial E}\right)_{\pi_{min}} = -\frac{q - q_0}{\Gamma RT} \tag{46}$$

Beyond the Henry concentration region, the sensitivity of measurement also depends on concentration c. This can easily be derived, e.g., from the mathematical formulation and the course of the adsorption isotherm, e.g., (24), (25), (31).

For the concentration dependence of the voltammetric current signal in the region of low concentrations the following approximate relation has been derived [14, 30, 78] by combining Eqs. (41), (42) and (31), or (24), (25)

$$I \sim k_2\Gamma_m\beta c - k_3\Gamma_m(\beta c)^2 \tag{47}$$

where k_2 and k_3 are combined constants.

The relative adsorptivity (β'/β'_{ref}) can thus be determined as $(I/I_{ref})_c$ for small values of *c*, provided that the measurements on the reference and measured systems were carried out under identical experimental conditions. An estimation of surface activity by comparing the magnitude of the measured signal with a reference substance - sodium dodecylbenzene sulphonate (NaDBS) or Triton X-100 - was used, e.g., in quantitation of the total content of SAS in solutions [14, 33, 78]. The concentration dependence *I-c* for desorption peaks of sodium dodecyl-sulphonate in 0.1M KCl is demonstrated in ***Fig.** 7*.

The values of the relative adsorptivity [14] of a series of oil products and further SAS were determined from analogous curves for solutions of NaDBS and Saratov oil.

An agreement between the relative data obtained from the *I-c* dependence and the relative values of *x* from electrocapillary measurements in the region of the potential of maximum adsorption has been confirmed.

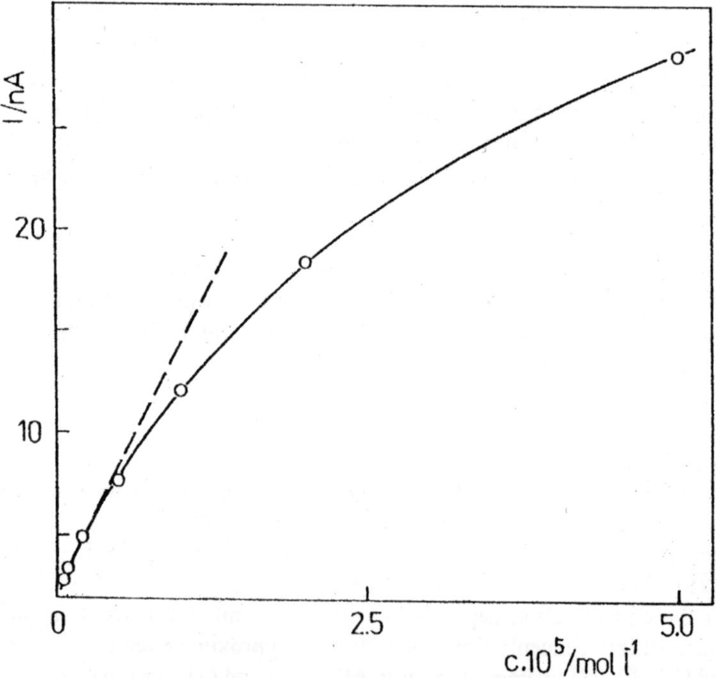

***Fig.** 7:* **Concentration dependence of current signal *I vs. c* for adsorption peaks of sodium dodecyl-sulphonate in 0.1M KCl**

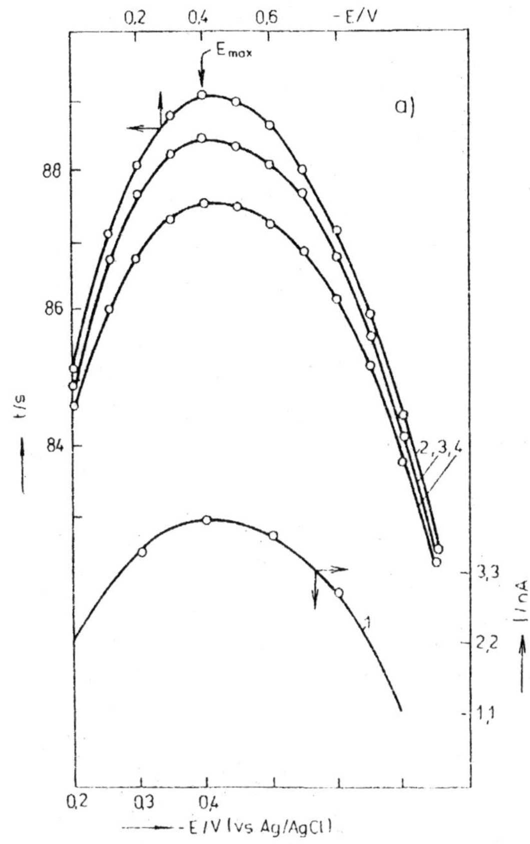

Fig. 8:
a) Comparison of the course of the peak-height *I* dependence on the accumulation potential E_{acc} (curve 1) with electro-capillary (adsorption) activity for trisbiphenyl (2,3,4 - curves *t-E*). After [40]
b) Potential dependence of the drop-time decrease Δt *vs.* *E* of the dropping mercury electrode
$5x10^{-5}$M dihydroxyphenylalanine (DOPA) in 1M KCl

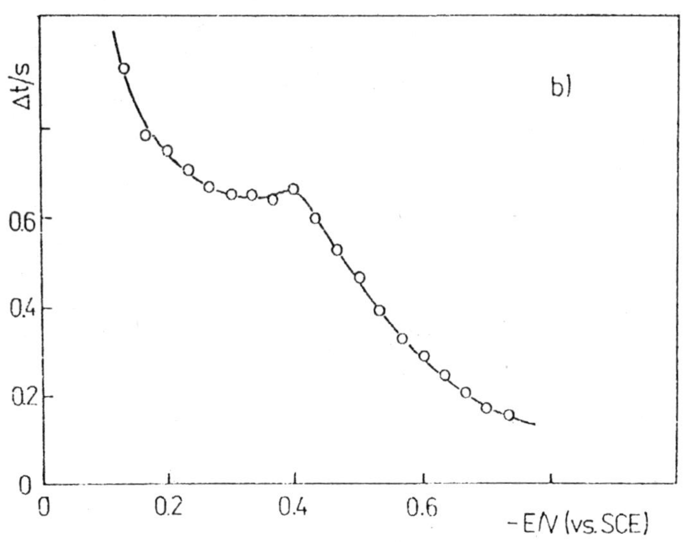

The highest sensitivity (and hence also the steepest *I-c* dependence for low c values) can be theoretically expected when the accumulation potential E_{acc} (i.e. the potential at which the adsorbing substance accumulates from the solution at the electrode, possibly under stirring) equals the potential of maximum adsorption E_{max}. For example, voltammetry of trichlorobiphenyl [84] (***Fig. 8a***), after adsorption accumulation in a buffer solution of pH 6.4 with 20 % methanol, yields a desorption peak at -1.1 V at renewed hanging mercury drop (*vs.* sat. Ag/AgCl electrode) which is highest if the accumulation is carried out at E_{acc} = -0.4 V. The I-E_{acc} [c = const] dependence agrees with the course of electrocapillary curves *t-E* and with the value of E_{max} determined from them. In practice we are more often confronted with non-parabolic, sometimes even quite unusual, courses of the I - E_{acc} or $\pi \sim \Delta t$ *vs.* E [c = const] dependences. Such is the case of the I-E_{acc} dependences of polyethylene glycols of various molecular weights [87] or potential dependence of the drop-time decrease Δt *vs.* E (where $\Delta t \sim \pi$) of the dropping mercury electrode due to the potential dependent adsorption of dihydroxyphenylalanin in 0.1M KCl (***Fig. 8b***). The reason for this can be, e.g., that the tensammetric signal is a non-linear function of Γ (or θ) in the region of high coverages, that the adsorption processes are affected by mercury dissolution in the positive potential region or by competitive adsorption, etc. Therefore, it pays to verify empirically the position of the expected accumulation potential, or of the potential of marked adsorption, before starting the measurements. The agreement of the maximum adsorption potential determined from electrocapillary curves with that obtained from the I-E_{acc} dependence is being confirmed in practice.

Medium effects, multicomponent adsorption, selectivity

The electrode surroundings have so far been considered as an inactive medium; nearest to this concept are the electrolytes $NaHSO_4$, Na_2CO_3, KF, NaF, NaOH. In reality, water as well as all electrolyte components including trace impurities interact to lesser or greater extent with the electrode. In order to maintain defined conditions, it is important to use freshly prepared and renewed electrodes. A transition from water to nonaqueous and mixed solvents brings about an increased surface activity of the supporting solution alone [88, 14]. In dependence on the concentration of individual solution components of different adsorptivity, a competitive adsorption occurs at the electrode surface as soon as it is dipped into the solution. In course of time the substances showing strongest adsorption prevail at the interface. A possible interference by ions of the supporting electrolyte (Cl^-, Br^-, I^-, etc.) can be demonstrated on reaction

$$O_{ads} + n_i(H_2O)_{ads} + \Sigma\, n_i(A_i)_{ads} = O_{des} + n_i(H_2O)_{des} + \Sigma\, n_i(A_i)_{des} \qquad (48)$$

where the indexes ads, des denote the state of adsorption and desorption, respectively. The Gibbs energy of adsorption, ΔG, is then an algebraic sum of the Gibbs energies [76] of individual reaction steps, i.e. of the salting-out contribution ΔG_S, of the interaction contributions of the components with the surface $[\Delta G_{Hg\text{-}org(H_2O);A_i}]$, and of the contribution $\Delta G'$ corresponding to mutual interactions of the particles in the adsorption layer. Equation

$$\Delta G = \Delta G_S + \Delta G_{Hg\text{-}org} + \Delta G_{Hg\text{-}H_2O} \pm \Delta G' + \Sigma\, \Delta G_{Hg\text{-}A_i} \qquad (49)$$

then expresses the Frumkin's adsorption isotherm (30), (31) the coefficients of which are average coefficients.

In a mixture of surfactants, the resulting activity either corresponds to the single components or fractions dominating in adsorptivity or to a combined effect of mutually competitive adsorption processes. If the studied substance is strongly adsorptive, the assumptions about a negligible effect of the surroundings are, as a rule, well justified. However, if the adsorption coefficient of the studied substance is not very high, the activity of other components of the solution can play a role, when $c \to 0$. Their adsorptivity, lower by several orders of magnitude, becomes then compensated by their concentration several orders of magnitude higher. As a consequence, there appear lags on the equilibrium curves of concentration dependences. The positive or negative surface activity of the supporting electrolyte can then be observed, as mentioned in the discussion of the properties of the coefficient $(\partial E/\partial \log a_{\pm})_{Ez}$. For example, ***Fig. 9*** shows the Δt-c dependence in 0.5M NaCl containing a small amount of motor oil, Saratov oil and an oil standard [21, 22]. However, the lags on the curves can also be caused by non-equilibrium conditions of adsorption or measurement, or by inter-molecular interactions in the double layer.

An increase in the concentration of a sufficiently pure supporting electrolyte [76] leads to an increase in the salting-out effect of the surfactant and, on the other hand, to a decrease in the thickness of the diffuse part of the double layer λ (***Fig. 1b***), and hence to a change in the potential gradient at the electrode surface which affects, among other factors, the value of Γ. The following relation holds,

$$\lambda \approx \sqrt{\frac{\varepsilon RT}{8\pi F^2 c}} \qquad (50)$$

where ε is the relative permittivity of the medium and F is the Faraday constant.

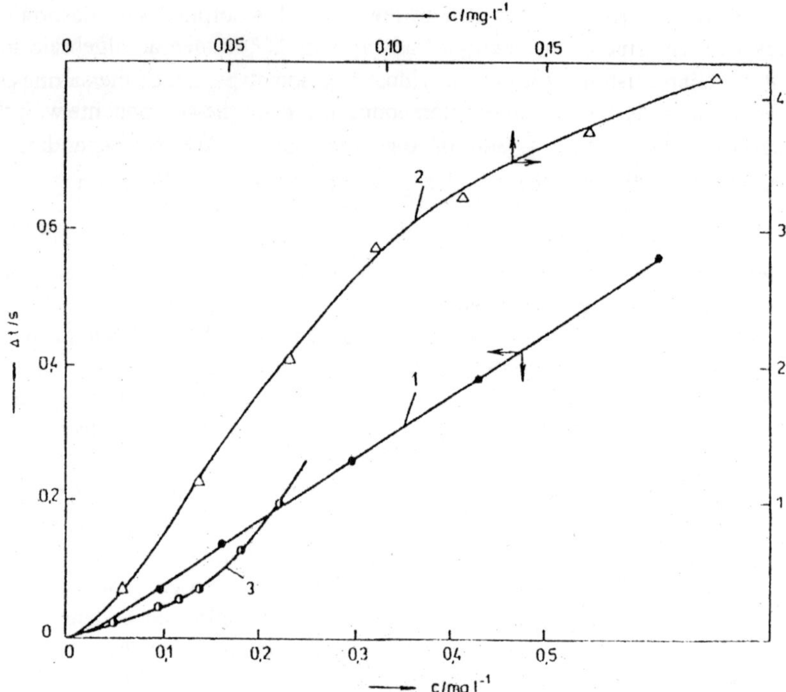

Fig. 9: Concentration dependence of drop-time shortening Δt as compared with drop-time in pure supporting electrolyte

t ≈ 93 s; solutions: Saratov oil (curve 1); motor oil (curve 2) in 0.5M NaCl, E = -0.4 V; oil standard in 0.5M NaCl, E = -0.6 V (curve 3); E *vs.* SCE

Reversible processes other than reversible adsorption can take place at the electrode surface, such as solvation-desolvation processes, ionic interactions etc. The pH of the solution is often an important factor affecting the AdSV results. It has been shown for 4,4'-bipyridyl adsorption [28] that a change in the pH can profoundly change the character of interfacial behaviour of a molecule and of its interaction with mercury surface.

The selectivity, or rather quasi-selectivity, of adsorption methods is based on specific shapes of the peaks [14, 22, 30], on the complex formation [8, 105], specific effects of surrounding factors on interfacial behaviour (effect of temperature change, of certain type of ions, etc.) [8, 31, 14], on the following of the character of the concentration dependence of the signal during gradual dilution of the sample [89], on parametric analysis of the course of concentration-, potential-, time- or another dependence of the measured signal {e.g., using relations (30-36) where the coefficients can characterize the studied system [14, 78]}, on analysis of Fourier-type signals [93], or on adjusting the sample after preliminary partial or complete separation [8, 90, 91]. In favourable cases it was possible to distinguish analytically two separate desorption peaks in a mixture of two substances differing markedly in their interfacial activity, e.g., when a fraction of polyethyleneglycol PEG 1500 can be determined in the presence of a PEG 4000 fraction [92].

The problem of selectivity of purely adsorptive measurements is still open [14]. One solution is sought in the field of modified surfaces [8]. Adsorption processes in which catalytic and/or other faradaic reactions take part in particular potential regions often provide favourable conditions for analytical utilization. Adsorption accumulation primarily increases the sensitivity of determination. When preparing an analytical procedure it is advisable to consider the possibility of combining faradaic reactions with adsorption, especially at higher coverages θ.

3.2.3 Process of establishing an adsorption equilibrium with respect to volume concentration

A gradual establishing of adsorption equilibrium can be experimentally followed when the rate of the determining processes is either comparable or less than the rate of the potential change, the rate of recording, the drop-time, etc. By far the most important controlling factor in analytical utilization of adsorption methods is the transport of the surfactant towards the electrode surface or away from it by diffusion, for sufficiently low concentrations (usually for $c \leq 10^{-5}$ - 10^{-4} mol l^{-1}). It precedes adsorption or follows desorption. It is then necessary to find the connection between the magnitude of the signal, the time of adsorption accumulation and concentration [14].

To establish these connections, it is necessary:

- to find an explicit relation between the measured signal and Γ;
- to determine the dependence of the surface concentration $\Gamma(t)$ on time t, measured from the moment of the surface renewal, for different volume concentrations c.

Considerable attention has been paid to the solution of the second problem. For a high adsorption coefficient, appropriately small concentration c and negligible concentration c_0 ($c_0 \ll c$) close to the electrode surface, the Koryta's equation [103] is valid up to the moment when $\Gamma(t)$ is about to reach Γ_m:

$$\Gamma(t)=\frac{2D^{1/2}c}{\pi^{1/2}}t^{1/2} \tag{51}$$

where D is the diffusion coefficient of the surfactant. For the region of validity of Henry's isotherm [$\Gamma(t) = \beta\Gamma_m c_0$] the following equation was derived [94]

$$\frac{\Gamma(t)}{\Gamma_e}=1-\exp\left[\frac{Dt}{(\Gamma_m\cdot\beta)^2}\right]\cdot erfc\left(\frac{D^{1/2}t^{1/2}}{\Gamma_m\cdot\beta}\right) \tag{52}$$

where $\Gamma_e = \beta\Gamma_m c$ and *erfc* is a tabulated function. More complicated cases, starting from the Langmuir isotherm, $\Gamma(t) = \beta\Gamma_m c_0(t)/[1 + \beta c_0(t)]$, were solved numerically and cannot be expressed in an explicit form. For analytical applications it would be desirable to find a relatively simple equation, e.g. of Koryta's type - Eq. (51), which would express an acceptable approximation of real dependences. A group of Japanese authors proposed, for evaluation of electrooxidation and electroreduction of flavine mononucleotide [95] and adriamycine [96] (connected with adsorption), to plot the dependence of the height of the a.c. polarographic peak (I_p) on the expression $\Gamma(t)$ for given concentration c, assuming that the following relation is valid over a wide time range,

$$I_p = k_4A\Gamma = k_5A\left[\frac{D}{r}t+2\left(\frac{Dt}{\pi}\right)^{1/2}\right]c \tag{53}$$

or, for c = const,

$$I_p = k_4'\Gamma(t) = k_5'\left[\frac{D}{r}t+2\left(\frac{Dt}{\pi}\right)^{1/2}\right] \tag{54}$$

where k and k' are constants and A denotes the area of the measuring electrode. The use of Eqs. (30-36) seems to be applicable to the fitting of the $I_p = I_p(t)$ dependence.

The problem of determination or verification of a quantitative relation between voltammetric signal I obtained by AdSV methods and surface concentration Γ for strongly adsorptive substances has not been satisfactorily solved, especially because of a lack of corresponding direct adsorption data. (The first direct equilibrium adsorption data for strongly adsorptive substances at very low concentrations were obtained in 1978 [11, 12].) It is usually assumed that I = const × $A\Gamma$. From a comparison of this assumption with Eq. (40) it is evident that it would be useful first to ascertain or verify the properties of the coefficient $(\partial q/\partial\Gamma)_E$ [12, 18] - cf. also Eq. (26), which is one of the basic coefficients in the theory of partial charge transfer in adsorbed state [16].

So far, we discussed free diffusion in a stationary solution. For the width of Nernst's diffusion layer δ at the electrode surface, the following expression holds

$$(\partial c/\partial x)_{x=0} = (c - c_0)/\delta \tag{55}$$

From here it follows then for the case $\Gamma(t) = \text{const} \times c_0(t)$ [94]

$$\Gamma(t)/\Gamma_e = 1 - \exp[-Dt/\Gamma_m\beta\delta] \tag{56}$$

(The value of δ is usually of the order of hundredths of a millimeter.)

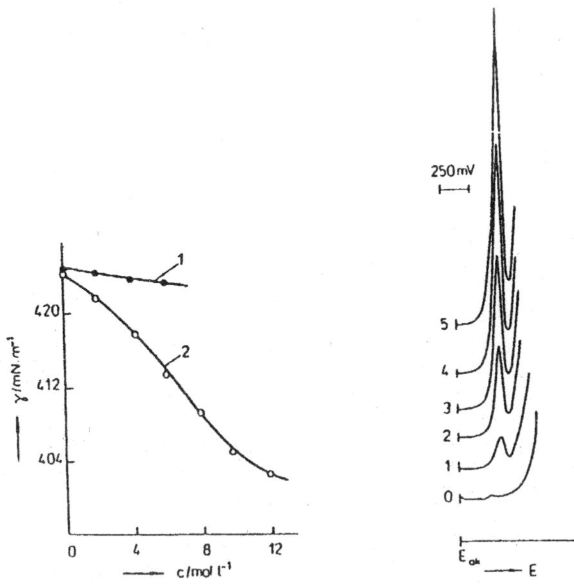

Fig. 10: a) Dependence of natural drop-time *t* on concentration *c* of solution of starch in 0.1 M KCl in a stationary solution (curve 1) and with convective accumulation (curve 2)

b) Effect of convective accumulation in AdSV; 10^{-6}M monensin in 0.1M NaCl; E_{acc} = -1.1 V *vs.* SCE

Each curve recorded with a new hanging drop, the numbers at the curves give the time of convective accumulation in minutes

The considerable effect of convective accumulation (with the solution being stirred [7, 12, 13]) consists in a marked thinning of the diffusion layer to $\delta' < \delta$, which remains constant under constant hydrodynamic conditions. As a consequence, in an equal time of accumulation t_{acc} the value of $\Gamma(t)$, and thereby also the sensitivity of measurement, markedly increases [54, 55]. ***Fig. 10a*** gives an example: drop-time $t \sim \Gamma$ for a freely dropping mercury electrode as a function of the concentration of starch in aqueous solution without and with convective accumulation (curves 1, 2). The effect of convective accumulation in AdSV is shown on the example of 10^{-6}M solution of monensin (***Fig. 10b***). Relations (7), (11), (12), (13) and other remain formally valid with the difference that instead of time-independent variables Γ, π, q, C and c, time-dependent ones appear, as $\Gamma(t)$, $\pi(t)$, etc.

Another attractive possibility of a considerable acceleration of the transport of matter to the electrode is now open by the application of microelectrodes including renewed mercury microelectrodes [43-49, 51] in the shape of a micro-drop or a mercury meniscus. A local increase in the surface concentration Γ can also be achieved by a controlled contraction of the electrode surface [19], the basis of the compression polarography, voltammetry and related techniques (see ***Fig. 11***), by means of ultrasound [54, 55], etc.

3.2.4 Irreversible Adsorption and Interfacial Processes

Irreversibility of interfacial processes can be easily evaluated by comparing AdSV records in a "pure" supporting electrolyte obtained with a renewed and nonrenewed mercury drop. As examples can serve a sensitive determination of thiourea and its derivatives in a perchlorate solution by cathodic stripping voltammetry utilizing chemisorption and formation of compounds with mercury [97, 98], AdSV method combined with the transfer of irreversibly adsorbed biomacromolecules at a mercury electrode [68, 106], determination of submicrogram amounts of nucleic acids [99, 107, 108] or determination of trace concentrations of anions or of substances forming products with mercury or mercury ions at positive potentials (Cl^- [1], dihydroxyphenyl-alanine [98], etc). Studied or analytically utilized can also be mutual interactions of adsorbed molecules as sucrose with Cl^- [100], reorientation of adsorbed layers [19, 22], association of mono- and oligonucleotides in adsorbed layer, condensation in the interphase [101], reorientation of films in presence of certain anions [28, 29, 102, 111], formation of ion pairs [28, 29] and a number of further effects described in publications on adsorption phenomena [14, 76, 69].

The sensitivity of interfacial methods towards molecular aggregates and micelles [14, 24] can be further analytically utilized for determining the so-called critical concentration of micelle formation c_{ccm} [66, 24], an important parameter for a given surfactant and medium. At the registered concentration

c_{ccm} (***Fig. 4d***) the dimensions of micelles are just comparable with the dimensions of the electrode double layer (Eq. (50)). The values of c_{ccm} are of the order of 10^{-5}-10^{-3} mol l^{-1}. For substances with an aliphatic carbon chain the c_{ccm} values depend approximately exponentially on the number of carbon atoms N, in agreement with the equation [67]

$$c_{ccm} = \exp(a - bN) \tag{57}$$

where a and b are constants.

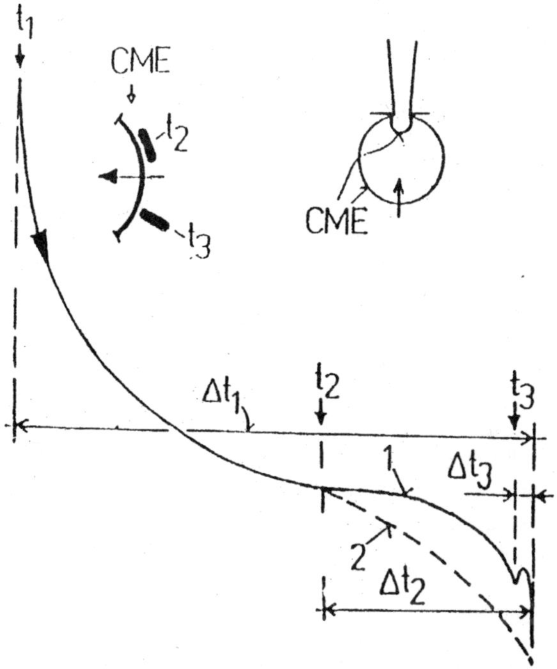

Fig. 11: **Current-time curves of Cd^{2+} reduction on the contracted mercury electrode (CME) in the presence of 1 × 10^{-5}M bipyridyl (BP), 5 × 10^{-5}M $CdSO_4$ in 0.1M H_2SO_4**

E = -0.65 V *vs.* SCE; curve 1 - recording in the presence of BP, compared with curve 2 - recording in the absence of BP

t_1 - start of the electrode surface contraction; equilibrium surface coverage of BP $\theta \approx 0{,}6$;

t_2 - reaching the monolayer coverage $\theta_a = 1$ of the planar orientation "α" of BP;

t_3 - reaching the monolayer coverage $\theta_b = 1$ of the perpendicular orientation "β" of BP;

$\Delta t_{1,2,3}$ - corresponding time ranges

Irreversible processes can also disturb analytical determinations. Their effect can usually be limited by finding more favourable working conditions or by changing the working regime.

3.3 Effect and selection of experimental conditions

The methods utilizing adsorption phenomena are more easily affected by the presence of undesired surfactants in the solution or by changes in the working procedure than methods based on purely faradaic processes; often there appear differences between the results obtained on a freshly renewed electrode and on one which was in use for some time. Nevertheless, the actual practical application of the methods of adsorption voltammetry and interfacial measurements is simple and near to other polarographic and voltammetric applications.

3.3.1 Temperature dependence

The temperature dependence of interfacial (electrocapillary) measurements of reversible systems follows in principle from a modified Gibbs-Lippmann equation [64, 65] analogous to Eq. (11):

$$d\pi = -\Delta S dT \approx -(Q_{ads}/T)\, dT \qquad (58)$$

valid for constant pressure and composition and for $Q_{ads} \approx Q_0(1 - \theta)$, where Q_{ads} is adsorption heat for given θ and Q_0 - adsorption heat for $\theta = 0$. The extent and polarity of the $d\pi$ changes depend on the magnitude and polarity of the constant Q_{ads}. The temperature coefficient $(\partial \Gamma / \partial T)_c = -0.17$ mN m^{-1} K^{-1} [76]; for pure electrolyte solutions it corresponds to Eq. (4). When the adsorption processes also involve adsorption kinetics, transport processes, faradaic reactions or irreversible phenomena, the resulting temperature dependence is also affected by these factors. The temperature change of the adsorption coefficient follows from Eqs. (11), (23) and (58).

In voltammetric measurements, the so far gathered values of the coefficient $(\partial i / \partial T)_c$ for "reversible" systems fall mostly within ± 1-3 % K^{-1}; they exceed these limits only exceptionally. More pronounced temperature dependences can be expected with irreversible interfacial processes (chemisorption) for which the absolute values of free enthalpy or of adsorption heats are considerably higher.

3.3.2 Electrode system

Although the described behaviour concerns various types of working electrodes (mercury or other liquids, gold, carbon, etc.) and electrode/liquid interfaces, its systematic study and verification were carried out above all thanks to the reproducibility and "ideal character" of the smooth surface of mercury electrodes, mainly in the stationary versions as HMDE or SMDE, in a specially adapted DME and, recently, also in the attractive miniaturized version mentioned below.

The basic assumption for a measuring stationary mercury electrode to be applicable to contemporary methods of adsorption voltammetry is, above all, its polarizability, renewability and a very good reproducibility. Of accessible commercial instruments, the laboratory bench apparatus SMDE [12, 34-39] proved best. The Czech concept of SMDE corresponds to the development stage of a new electrode system based on a mercury mini-, semimicro- and microelectrode [12, 14, 35, 37-39, 41, 43-51]. Compared with the SMDE, the latter electrode offers the following additional advantages :

- Introduction of a wide variety of new miniaturized electrode regimes: static or stepwise growing mercury drop and meniscus microelectrodes μE (denoted as DMμE,, HMμE, SMμE);
- A substantially greater range of functional parameters (e.g., sizes from radius r = 15 μm to the usual r < 1 mm), reproducibility of DME to ± 0.001 %; of HMDE to ± 0.1 - 0.2 %; of renewed mercury E to ± 1 - 3 %; a good stability even under strong vibrations, etc.;
- Research and utilization of new possibilities offered by new types of capillary electrodes for general and special purposes, by capillary modules, by integrated microelectrode systems, by "absolute" measurements without calibration, based on the reproducibility of individually prepared capillary and meniscus electrodes, etc.

The apparatus (***Fig.12a***; 2a) is miniaturized, portable, applicable in laboratory as well as under operating conditions. It utilizes miniaturized pen-type mercury electrodes (***Fig. 13***). Depending on the selected type of capillary it enables measurements in regimes usual for SMDE as well as in various modes for producing renewed mercury microelectrodes.

Mercury microelectrodes (***Fig. 12***) permit a markedly accelerated transport of substances toward the electrode, an increase in the current density under good signal/noise ratio, speeding up of the analytical procedure, analyses in media with a low conductivity and in the presence of oxygen, new accumulation conditions, ways of polarization, use of simple instruments. ***Fig. 12b*** brings a comparison of results obtained with miniaturized mercury electrodes of different sizes. The smaller is the electrode, the faster is the equilibrium coverage attained and the higher is the current density of the AdSV signal. ***Fig. 12c*** ilustrates a steady state using the meniscus micro-electrode HMμE at r = 20 mm.

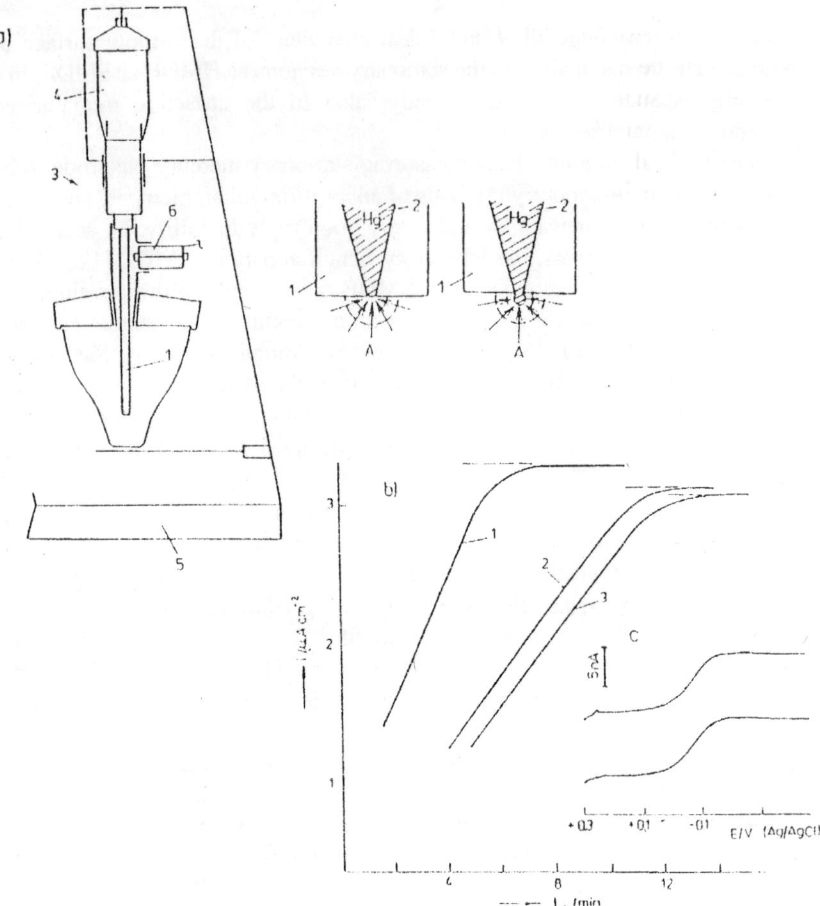

Fig. 12 **a) Scheme of the mercury microdrop or micromeniscus electrode**

A - spherical conditions of transport

1 - capillary, 2 - mercury, 3 - electrode with control mechanism (4) and with programmable pulse source (5); 6 - minihammer for drop detachment, radius of the mercury meniscus ~ 30 μm

b) Dependence of current density $i = I/A$ (A - electrode surface area) on time of accumulation t_{acc} under spontaneous diffusion

Solution-about 3×10^{-6} mol l^{-1} monensin in 0.1M NaCl; approximate volume of the microdrop: 1 - 2×10^{-3} ml; 2 - 8×10^{-3} ml; 3 - 3×10^{-2} ml

c) Steady-state voltammogram obtained for oxidation of 10^{-3}M ferrocene in acetonitrile (0.05M Et_4NClO_4) at the renewed mercury meniscus microelectrode of radius $r = 20$ μm; scan rate 10 mV s^{-1}

For electrocapillary measurements, a dropping mercury electrode made from a spindle capillary [9, 35, 62, 47-51], with drop-time between 35 and 130 s is best suited.

From the practical point of view, solid (carbon, gold, platinum, etc.) electrodes and microelectrodes play extraordinarily significant role for electroanalysis [3, 8].

3.3.3 Composition of solutions, measuring conditions, operation methods

The experimental arrangement and the working regime in AdSV correspond to usual conditions in polarography of organic compounds [3, 4]. However, greater attention must be paid to establishing and maintaining a strict operating procedure (potential and time of accumulation, rate of potential scanning, composition of basic solution, etc.), to selection of a sufficiently inactive supporting electrolyte, to purity of mercury, etc. The presence of a small amount of oxygen does not usually severely disturb the measurements. Thanks to thermal inertia of the medium in ordinary laboratories it is not essential to use thermostatted cells. A more profound effect on the results can be exerted by a change in the pH of the solution. It is necessary to check cleanliness of the capillary (especially of its mouth), let few drops fall before each measurement, polarize the electrode several times within the required potential range in a stirred as well as unstirred solution, keep the working electrodes as often as possible under controlled applied potential, etc. In case of electrocapillary measurements it is important to isolate the cell from accidental vibrations, e.g. due to the stirrer or the thermostat.

3.4 Conclusions

The methods of adsorption voltammetry and interfacial measurements are now going through a period of practical and theoretical development and of the search for new analytical applications. The already acquired knowledge of interfacial processes on the boundary mercury electrode-solution is being utilized and widened. Some actual theoretical and practical problems connected with this relatively new field of scientific research have been outlined in the present article.

3.5 References

1. Heyrovský J., Kůta J.: *Principles of Polarography*. NSAV, Prague 1965.
2. Kalvoda R., Parsons R.: *Electrochemistry, Research and Development*. Plenum Press, New York, London 1985.
3. Galus Z.: *Fundamentals of Electrochemical Analysis*. E. Horwood, Warszaw 1994.
4. Bond. A. M.: *Modern Polarographic Methods in Analytical Chemistry*. M. Dekker, New York 1980.
5. Vydra F., Štulík K., Juláková E.: *Electrochemical Stripping Analysis*. E. Horwood, New York 1976.
6. Wang. J.: *Electroanalytical Techniques in Clinical Chemistry and Laboratory Medicine*. VCH Publishers, Inc., New York 1988.
7. Kalvoda R.: Anal. Chim. Acta *138*, 11 (1982).
8. Kalvoda R.: Fresenius J. Anal. Chem. *349*, 565 (1994).
9. Kalvoda R., Budnikov G. K.: Collect. Czech. Chem. Commun. *28*, 838 (1963).
10. Kalvoda R.: J. Electroanal. Chem. *180*, 307 (1984); Ann. Chim. (Rome) *73*, 239 (1983).
11. Novotný L.: Czech. pat. AO 210 852, PV-9200, Prague 1978.
12. Novotný L.: PhD Thesis. J. Heyrovský Institute, Prague 1980/81.
13. Novotný L., Smoler I.: J. Electroanal. Chem. *146*, 183 (1983).
14. Novotný L.: in *New Trends in Analytical Chemistry* (Zýka J., Ed.), (in Czech). Publishers of Technical Literature, Prague 1989. P. 56-87.
15. Kopanica M.: in *Advanced Instrumental Methods of Chemical Analysis*, (Churáček J., Ed.). Academia, Prague 1993. P.75-95.
16. Jehring H.: *Elektrosorptionanalyse*. Akademieverlag, Berlin 1975.
17. Novotný L., Smoler I.: Collect. Czech. Chem. Commun. *50*, 2525 (1985); Proc. of the J. Heyrovský Mem. Congress, Vol.II, Prague 1980. P. 128.
18. Novotný L., Smoler I., Kůta J.: Collect. Czech. Chem. Commun. *48*, 964 (1983).
19. Novotný L.: Electroanalysis, submitted.
20. Vytřas K.: in *Advanced Instrumental Methods of Chemical Analysis* (Churáček J., Ed.). Academia, Prague 1993. P.142-158.
21. Novotný L., Kalvoda R.: Collect. Czech. Chem. Commun. *51*, 1595 (1986).
22. Kalvoda R., Novotný L.: Collect. Czech. Chem. Commun. *51*, 1587 (1986).
23. Samec Z., Mareček V.: J. Electroanal. Chem. *200*, 17 (1986); *333*, 319 (1992).
24. Rudovský J., Novotný L.: Vod. Hosp. *B 4*, 97 (1984).
25. Kalvoda R., Novotný L.: Vod. Hosp. *B 11*, 291 (1984); Collect. Czech. Chem. Commun. *51*, 1587 (1986).
26. Kalvoda R.: Anal. Chim. Acta *162* , 197 (1984).
27. Kalvoda R.: Ann. Chim. (Rome) *73*, 239 (1983).
28. Heyrovský M., Novotný L.: Collect. Czech. Chem. Commun. *52*, 56 (1987).
29. Heyrovský M., Novotný L.: Collect. Czech. Chem. Commun. *52*, 1097 (1987)
30. Kováč M., Kalvoda R., Novotný L.: Electroanalysis *5*, 171 (1993).
31. Krumbein A., Novotný L., Retter U.: Collect. Czech. Chem. Commun. *59*, 1745 (1994).
32. Wandlowski Th., Heyrovský M., Novotný L.: Electrochim. Acta *37*, No. 14, 2663 (1992).
33. Novotný L., Navrátil T.: Chem. Listy, submitted.

34. Peterson W. M.: Amer. Lab. *112*, 51 (1980).
35. Novotný L.: Czech.pat. AO 202 316; 202 772 (1978); 220 439 (1982); Proc. of the J. Heyrovský Mem. Congress, Vol.II, p.129, Prague 1980.
36. Novotný L.: in *User's Guide for the Polarographic Analyzer PA 3*, Laboratory Instruments, Prague 1982.
37. Novotný L.: Czech.pat. AO 219 475, PV 1995-80, Prague 1980.
38. Novotný L.: Czech.pat. AO 224 754, PV 8383-81, Prague 1981.
39. Novotný L.: Czech.pat. AO 223 626, PV 8449-81, Prague 1981.
40. Trojánek A., Holub I., Křesťan L., Novotný L.: Chem. Listy *75*, 1091 (1981).
41. Novotný L.: Proc. of the "Analytiktreffen 1985", Neubrandenburg, Nov. 10-14th, 1985. P. 59.
42. Heyrovský M., Novotný L., Smoler I.: in *Electrochemistry, Past and Present,* (Stock J. T. and Orna M. V., Eds.). American Chemical Society, Washington 1989. P. 370-379.
43. Novotný L.: Proc. of International Conf. on "New Trends in Polarography and Related Techniques", Fermo (Italy) 1986. P. 41-43,
44. Novotný L.: Czech. pat. appl. V-22864/89, Prague 1989.
45. Novotný L.: Proc. of the J. Heyrovský Centennial Congr. on Polarography, *180*, Prague 1990. P. Tu-105.
46. Novotný L.: *Concept of new type of electrode systems* (*UMME, UMmE, UMμE*), ESEAC, The Netherlands, Noordwijkerhout 1992.
47. Novotný L.: Europat. Appl. No 91101283.9 (1990).
48. Novotný L.: US Pat. No 5 294 324; priority 1990.
49. Novotný L.: US Pat. No 5 173 101; priority 1990.
50. Novotný L., Kalvoda R., Novák V., Uher K.: Europat. No 376920 (1990); priority 1988.
51. Novotný L.: Electroanalysis *2*, 257 (1990).
52. Novotný L., Herout M.: Czech. pat. AO 220 125, Prague 1989.
53. Novotný L.: Czech. pat. (submitted).
54. Novotný L.: Czech. pat. PV 2933-93, Prague 1993.
55. Novotný L.: Chem. Listy, accepted, *89*, Prague 1995.
56. Novotný L.: Collect. Czech. Chem. Commun. *50*, 2517 (1985); AO 210 894, PV 7544-79, Prague 1979.
57. Novotný L., Smoler I., Kůta J.: Czech. pat. PV 185 983, Prague 1976.
58. Novotný L.: Czech.pat. AO 181 889, PV 6250-76, Prague 1976.
59. Novotný L.: Czech.pat. AO 223 292, PV 8792-81, Prague 1981.
60. Novotný L., Smoler I., Kůta J.: J. Electroanal. Chem. *88*, 161 (1978).
61. Novotný L., Večerník J.: Czech. pat. AO 201 724, Prague 1980.
62. Novotný L.: Czech.pat.AO 207 870, PV 7545/6-79; AO 207 871, PV 7584-79, Prague 1979.
63. Novotný L.: Czech. pat. PV 9112-88, Prague 1988.
64. Damaskin B.B., Petrii O.A., Batrakov V. V.: *Adsorption of Organic Compounds on Electrodes*. Plenum Press, New York 1971.
65. Koryta J., Dvořák J.: *Principles of Electrochemistry*, J. Wiley and Sons, New York 1987.
66. Adamson A. W.: *Physical Chemistry of Surfaces*. New York 1967.

67. Tamaru K.: *Kapilyarnaya Khimiya*. Mir, Moscow 1983.
68. Paleček E., Postbieglová I.: J. Electroanal. Chem. *214*, 359 (1986).
69. Dryhurst G.: *Electrochemistry of Biological Molecules*. Academic Press, New York, 1977.
70. Novotný L.: Czech. pat. PV 4202-91, Prague 1991.
71. Novotný L., Hönig J., Herout M.: ESEAC, The Netherlands, Noordwijkerhout 1992.
72. Novotný L., Kopanica M., Kalvoda R., Šestáková I., Navrátil T., Jakubetzová A., Heyrovský M., Stará V.: ESEAC, The Netherlands, Noordwijkerhout 1992.
73. Novotný L.: Vod. Hosp. *9*, 16 (1994).
74. Novotný L., Heyrovský M.: Trends in Anal. Chem. *6*, No. 7, 176 (1987).
75. Ménard H., Kimmerle F. M.: J. Electroanal. Chem. *47*, 375 (1973).
76. Damaskin B.B., Petrii O. A.: *Vvedeniye v Elektrokhimicheskuyu Kinetiku*. Vyshchaya shkola, Moskva 1983.
77. Delahay P.: *Double Layer and Electrode Kinetics*. Publishers, Inc., New York 1966.
78. Novotný L.: in preparation for publication.
79. Novotný L.: Czech. pat., submitted.
80. Novotný L.: in *Software for PC-ETP (Polarograph)*, Polaro-Sensors, Prague 1995.
81. Mohilner D. M.: in *Electroanalytical Chemistry* (Bard A. J.,, Ed.), Vol.I. M. Dekker, New York 1966. P. 289.
82. Kůta J., Smoler I.: Collect. Czech. Chem. Commun. *40,* 225 (1975)
83. Behr B., Drogowska M.: J. Electroanal. Chem. *82,* 317 (1977).
84. Lam N. K., Kopanica M.: Anal. Chim. Acta *161,* 315 (1984).84.
85. Novotný L.: in *Principles of Electrochemistry* (J. Koryta, J. Dvořák, Eds.). J. Wiley and Sons, p. 217, New York 1987.
86. Frumkin A. N., Damaskin B. B.: J. Electroanal. Chem. *3*, 36 (1962).
87. Batycka H., Lukaczewski Z.: Anal. Chim. Acta *162,* 207 (1984).
88. Borkowska Z., Dojlido J.: Roczn. Chem. *52,* 2007 (1978).
89. Novotný L.: Czech. pat. PV 1433-93, PV 1215-95, Prague 1993, 1995.
90. Štulík K., Pacáková V.: *Electroanalytical Measurements in Flowing Liquids*. Ellis Horwood, Chichester 1987.
91. Šestáková I., Kopanica M.: Talanta *35*, No. 10, 816 (1988).
92. Batycka H., Lukaszewski Z.: Anal. Chim. Acta *162*, 215 (1984).
93. Novotný L.: Czech. pat., submitted, Prague 1995.
94. Delahay P., Trachtenberg I.: J. Amer. Chem. Soc. *79,* 2355 (1957); *80,* 2628 (1958).
95. Kakutani T., Kano K., Ando S., Senda M.: Bull. Chem. Soc. Jap. *54*, 884 (1981).
96. Kano K., Konse T., Nishimura N., Kubota T.: Bull. Chem. Soc. Jap. *57*, 2383 (1984).
97. Stará V., Kopanica M.: Anal. Chim.Acta *159,* 105 (1984).
98. Kalvoda R.: J. Electroanal. Chem. *214*, 191 (1986).
99. Paleček E., Boublíková P., Jelen F.: Anal. Chim. Acta *187*, 99 (1986).
100. Krishman M., de Levie R.: J. Electroanal. Chem. *131,* 97 (1982).
101. Sridharan R., de Levie R.: J. Electroanal. Chem. *201,* 133 (1986).
102. Hills G., Silva F.: J. Electroanal. Chem. *137,* 387 (1982).
103. Koryta J.: Collect. Czech.Chem. Commun. *18*, 206 (1953).
104. Batina N., Kozarac Z., Cocovic B.: J. Electroanal. Chem. 188,, 153 (1985).
105. Lam N. K., Kalvoda R., Kopanica M.: Anal. Chim. Acta 154, 79 (1983).

106. Heyrovský M., Mader P.,Veselá V., Fedurco M., Vavřička S.: J. Electroanal. Chem *369*, 53 (1994); *375*, 371 (1994).
107. Paleček E.: Bioelectrochem. Bioenergetics *20*, 179 (1988)
108. Teijero C., Nejedlý K., Paleček E.: J. Biomol. Struct. Dynamics *11*, No. 2, 313 (1993).
109. Bersier P. M., Howell J., Bruntlett C.: Analyst *119*, 219 (1994).
110. Wang J.: Analyst *119*, 763 (1994).
111. Vetterl V., de Levie R.: J. Electroanal. Chem. *310*, 305 (1991).

4

ADSORPTIVE STRIPPING VOLTAMMETRY

Robert Kalvoda

Abstract

The techniques of adsorptive stripping voltammetry and its application in environmental analysis are described. Properties as the sensitivity, experimental arrangement, or elimination of various types of interferences are assessed.

Key Words

voltammetry, adsorptive stripping voltammetry, polarography, environmental analysis, trace analysis, heavy metal analysis.

4.1 Introduction

Most applications of chemical analysis to environmental protection involve trace determinations, often at a part-per-billion level or lower. Among methods that can satisfy this demand belongs, without doubt, polarography/voltammetry. However, to attain its today's microanalytical quality the original Heyrovský method had to pass through improvements and modifications, directed mainly to a substantial improvement in its sensitivity.

In the first place there are instrumental approaches toward suppression of undesirable, but inevitablee charging current and thus toward an improvement of the signal-to-noise ratio. The various modes of pulse methods can be regarded as one of the successful results.

In addition to instrumental methods, simple and cheap laboratory techniques have been developed to ensure lowest possible detection limits in polarographic/voltammetric analysis. These methods are based on accumulation of the compound to be determined at the electrode surface, followed by its voltammetric determination. In this sense, the most popular method which has gained worldwide popularity, is anodic stripping voltammetry (ASV), used mainly in trace analysis of heavy toxic metals. This method makes use of electrolytic deposition of traces of metal ions at the surface of the working electrode. In addition, some anions or organic compounds can be accumulated at a mercury electrode on the basis of the formation of insoluble compounds with the mercury ions produced by anodic dissolution of the mercury electrode. In this type of cathodic stripping voltammetry (CVS), the reduction process of the thus formed mercury compound is then followed.

Along with the previously described electrolytic processes, some other principles can be used for accumulation of the compound to be determined. One of these is adsorption. To this finding we came at the Heyrovský Institute during our studies connected with alternating current oscillographic polarography. Such an adsorptive accumulation of a depolarizer at a mercury electrode leading to an increase in the sensitivity of oscillographic polarography was observed in the early fifties, e.g., for a solution of elemental sulphur [1], mercurous thiosulphate, poorly soluble inorganic substances, etc. [2]. In fact, this mode of polarography, where the electrode is polarized by sinusoidal alternating current and the electrode potential or, better, its differential function, $dE/dt = f(E)$, is measured, yields good possibilities for studies of adsorption phenomena. An example of adsorptive accumulation of a reducible organic compound (e.g. acetophenone) at a constant voltage prior to electrode polarization by a single alternating current cycle is given in [3]. The adsorptive accumulation of the reduction product of methylene blue at a hanging mercury drop electrode has also been described by the Czech school [4].

The adsorptive accumulation of substances at electrodes began to be exploited in practical analysis towards the end of the 1980's, as improved types of hanging mercury drop electrodes appeared on the market.

In the following paragraphs, some fundamentals and properties of AdSV, such as the theoretical background, experimental arrangement and possibilities in trace analysis, are discussed. Prospects of this method are also outlined.

4.2 Theoretical background

Many, mostly organic compounds exhibit surface-active properties that are manifested by their adsorption from solution onto the surface of a solid phase. This phenomenon forms the basis for adsorptive stripping voltammetry (AdSV). The action of interfacial forces at the boundary between two phases, in this case an electrode in solution, leads to the formation of an interface with a thickness of molecular dimensions. If these interfacial forces lead to an increase in the concentration of some substances at the solid phase-solution interface compared to the concentration in the solution, then this substance is adsorbed on the surface of the solid phase. Adsorption equilibrium is attained between the substance present in the solution and at the surface of the electrode. At a given temperature, the amount of substance adsorbed is dependent on the concentration in the solution. The velocity of formation of the adsorbed layer is affected both by the rate of the actual adsorption of the substance from the solution layer in direct contact with the electrode and also by the rate of diffusion of the substance from the bulk of the solution to the electrode surface. The slower of these processes then becomes the rate-controlling step in the formation of the adsorbate.

For fast, diffusion controlled adsorption, which occurs most often in adsorptive accumulation, assuming that mass is transported under the limiting current conditions, the following relationship has been derived [5] for the peak current of a compound at a HMDE, provided that the peak current is a function of the surface concentration

$$I_p = kAG \tag{1}$$

$$I_p = kAC/(D/r)t_{acc} + 2(D/\pi)^{1/2}t_{acc}^{1/2} \tag{2}$$

where k is a proportionality constant, A the electrode surface area, Γ is the surface concentration of the compound and C its concentration, D is its diffusion coefficient, r is the HMDE radius and t_{acc} is the accumulation period. For large values of C and/or t_{acc}, I_p approaches a limiting value of I_p^{max},

$$I_p^{max} = kA\Gamma_m \tag{3}$$

For complete coverage, Γ_m, Eq.(4) has been derived by Koryta [6]:

$$\Gamma_m = 7{,}36 \times 10^{-4}\ CD^{1/2}t^{1/2} \tag{4}$$

where t is the time required for establishment of complete electrode coverage. It has been found that I_p increases linearly with $t_{acc}^{1/2}$ [7-9] (of course, assuming that the coverage is smaller than 1, there is no interaction among the adsorbed molecules and the adsorption of the compound is controlled by its diffusion to the electrode); this is a criterion for diffusion-controlled adsorption. The I_p value is roughly proportional to the product of C and $t_{acc}^{1/2}$, provided that neither of these values is too large. The linear dependence of I_p on the scan rate v is predicted by the equation:

$$I_p = n^2F^2\Gamma v/4RT \tag{5}$$

derived by Brown and Anson [10]. This is a proof of adsorption, in contrast to the $v^{1/2}$ dependence valid for pure Faradaic processes. Concerning adsorption parameters, for the concentration dependence of the voltammetric signal in the region of small concentration levels, Novotný [11] derived the equation

$$I = k_2\Gamma_m\beta C - k_3\Gamma_m\beta C^2 \tag{6}$$

and, thus, for the slope I/C

$$I/C = k_2\Gamma_m\beta - k_3\Gamma_m\beta C \tag{7}$$

where k_2 and k_3 are proportionality constants and Γ_m denotes the maximum coverage of the electrode. The slope I/C can than be considered as a function of the adsorption coefficient β,

$$I/C = f(\beta) \tag{8}$$

and the adsorptivity, expressed by this slope, as a parameter characteristic for the compound under study.

An equation has been derived by Wenrui Jin [12] for the dependence of I_p on parameters such as the analyte concentration and the accumulation time for application of microelectrodes. These electrodes, due to their large edge effect, exhibit uniquely enhanced mass transport characteristics and a lower *iR* drop. It has been found that the dependences of the peak current on the voltage scan rate, accumulation time in a quiescent solution, the mercury ultramicroelectrode radius and the bulk concentration of adsorbed substances, are in agreement with theoretical prediction. Eqs. (9) and (10) thus hold for reversible and irreversible processes, respectively [12]

$$I_p = (n^2F^2/4RT)ADvt_{acc}C/r \tag{9}$$

$$I_p = (n^2F^2/4eRT)\alpha ADvt_{acc}C/r \tag{10}$$

where e is the base of natural logarithms, r is the ultramicroelectrode radius, v is the scan rate and α the charge-transfer coefficient.

If the amount of substance adsorbed on the electrode surface is determined by the rate of the adsorption process, which is smaller than the diffusion rate, it can be assumed that the concentration of the surface active compound at the electrode surface is equal to its concentration in solution. Similar conditions are valid for weak adsorption of a surface active substance. Neither of these cases is of interest for AdSV.

It is thus evident that the conditions in AdSV are rather complicated from the theoretical point of view, as many factors concerning adsorption of the species on the electrode surface must be taken into account. These include the type of adsorption, diffusion conditions and the type of electrode reaction of the adsorbed species. This reaction may be an electrolytic process or it can only be adsorption/desorption, such as the case with an electrochemically inactive compound yielding tensammetric peaks. In this latter case it is necessary that the adsorption coefficient of the substance in the given supporting electrolyte has a value of about 10^3 l mol^{-1} or greater, roughly corresponding to the reciprocal of the concentration for a 50% coverage of the electrode by the adsorbed substance.

It follows from the above discussion that AdSV can be utilized for determination of substances that are readily adsorbed at an electrode. Here also belong compounds characterized by a poor solubility in aqueous supporting electrolytes, e.g. higher aliphatic alcohols, higher fatty acids, aromatic hydrocarbons, aromatic nitrocompounds, compounds with condensed rings, alkaloids, antibiotics, detergents and many others. If the molecules of these compounds contain a group that can undergo a faradaic process at the electrode, the voltammetric curve corresponds to the reduction or oxidation of the adsorbed substance. For electroinactive substances, only a tensammetric peak is formed on the curve in the region corresponding to the desorption of the accumulated substance. It has been found for substances of this type that an accumulation effect can be expected when the surface active substance yields well-developed tensammetric peaks at a dropping mercury electrode at concentrations of about 10^{-5} mol l^{-1}. The greater the adsorption coefficient, the higher and narrower is the tensammetric peak. The adsorption can also be affected by suitable choice of the supporting electrolyte solution and is often accompanied by a salting-out effect. The AdSV method is also useful for poorly soluble inorganic compounds, especially metal complexes (see below).

4.3 Results and discussion

4.3.1 Experimental arrangement and working conditions

4.3.1.1 APPARATUS AND ELECTRODES

The AdSV procedure can be carried out with commercial polarographic instruments, using both the classical polarographic method - i.e., d.c. polarography and pulse methods, especially differential-pulse voltammetry and square-wave voltammetry. In choosing a suitable voltammetric method, it should be borne in mind that the measurement of the peak height can often be complicated by unfavourable shapes of the background curves, especially at more positive potentials. The d.c. method is then preferable. A computer controlled apparatus with automatic timing of the individual operations is useful for controlling the individual steps in AdSV measurements, i.e. the accumulation time, solution stirring, duration of the rest period, and initiation of polarization.

Most types of electrodes used in voltammetry can be employed in AdSV provided that a constant, completely reproducible surface can be ensured throughout the whole measurement or, better, during a series of measurements. The above requirements on the reproducibility can best be satisfied by a hanging mercury drop electrode. The carbon paste electrodes or platinum electrodes are mostly used for measurements, where the adsorbed compound is oxidized during the voltammetric scan, because they can be polarized to more positive potentials than mercury. AdSV can be carried out with chemically modified electrodes which consist of an electrode material to the surface of which substances or functional groups are bound that alter their properties, primarily improving the sensitivity and selectivity. The modifier can also be used as a filter preventing passage of interferents to the electrode surface (for more details on chemically modified electrodes, see [13]). However, their use in routine analysis can often be complicated due to difficulties in the reproduction of the electrode surface; moreover, the accumulation process at these electrodes is far more complex than for reversible adsorption at a mercury electrode.

4.3.1.2 GENERAL EXAMINATION OF THE POSSIBILITIES OF THE AdSV DETERMINATIONS

It is relatively simple to decide whether a substance can be determined by AdSV at a mercury electrode. First, the differential-pulse or square-wave voltammetric behaviour of the compound (at a concentration of cca 10^{-6} mol dm^{-3}) is examined at a hanging mercury drop electrode in various supporting electrolytes. In an optimum supporting electrolyte, the initial potential is then set to 0 V or -0.1 V *vs.* SCE, a new mercury drop is formed and the voltage scan towards more negative potentials is immediately begun at a rate of cca 20 mV s^{-1}. After the voltammetric curve has been recorded, a new mercury drop is again formed and the same initial potential is applied for a period of cca 60 s to the working electrode in stirred solution. Following this accumulation period (A_{acc}), the stirring is stopped and the voltage is scaned as previously, after a rest period of cca 10 s. If the examined compound is accumulated, then a substantial increase in the peak current is observed, as not only the substance transported to the electrode by diffusion, but also that adsorbed on the electrode surface is reduced during the voltage scan.

For oxidizable organic compounds, a solid working electrode is used in a similar way: the accumulation is studied at 0 V or with an „open circuit“ and then the voltammetric curve is recorded toward more positive potentials. With solid electrodes, accumulation often occurs simply on immersing the electrode in a stirred solution for a certain time, t_{acc}. The electrode is then rinsed and transferred to a „pure" supporting electrolyte, where the actual voltammetric determination is carried out. This procedure has a certain advantage in that the effects of interferents in the sample (e.g. compounds yielding peaks close to the analyte peak) can thus be eliminated. However, interfering substances can be adsorbed from the sample during the accumulation period and can sometimes greatly influence the AdSV determination.

After these preliminary investigations, the most suitable accumulation potential E_{acc} is found by examining the dependence of the peak current I_p on E_{acc}. If chemisorption participates in the adsorption process, a more positive E_{acc} value must often be employed, e.g. +0.1 V. This is especially true for cathodic stripping voltammetry where the electrode product is also adsorbed. It is often useful in analytical applications to employ a more positive E_{acc} value so that traces of heavy metal impurities are not deposited at the electrode. The optimal accumulation time must also be found. The peak height increases linearly with increasing t_{acc} up to a value corresponding to complete coverage of the electrode by adsorbate or to the attainment of equilibrium between the compound adsorbed on the electrode surface and in the bulk of the solution. Complete coverage is attained faster in stirred solutions. The t_{acc} value at which the limiting I_p value is attained also depends on the sample concentration.

The dependence of I_p on the analyte concentration should be linear over a reasonably wide range. The method of standard additions can be used for quantitative measurements. Three additions of a standard solution are recommended to ensure that the measured I_p values correspond to the linear part of the calibration curve. When the I_p value does not increase linearly during the standard additions, the sample solution must be diluted or a shorter accumulation time employed. In such a case, accumulation can also be performed in unstirred solutions.

It is recommended that a blank accumulation experiment be carried out in the pure supporting electrolyte, especially for longer t_{acc} values, because surface active impurities in the solution can also be adsorbed on the electrode and yield parasitic peaks, or even affect the accumulation process of the analyte due to competitive adsorption. Inhibitive effects from such competitive action can be avoided by using shorter t_{acc} times (e.g., 15 to 30 s).

4.3.2 Applications in environmental analysis

As mentioned above, environmental analysis is mostly applied trace analysis - and AdSV is ideally suited for trace analysis, as detection limits for electroactive compounds are in the range of 10^{-10} mol dm^{-3} concentration.

However, this value can only be attained under ideal conditions, which rarely occur in practice. The main factor limiting the sensitivity and sometimes also the applicability of this method, is competitive adsorption of other surface-active substances present in the test solution. The analyte peak height then decreases or the peak disappears completely. Therefore, practical detection limits can be expected to be in the concentration range from 1×10^{-8} to 1×10^{-9} mol dm^{-3}.

The scope of application of AdSV ranges from metal trace analysis to analysis of organic compounds and biochemical applications.

4.3.2.1 TRACE METAL DETERMINATIONS

Most published papers deal with metal trace analysis, exploiting the fact that many metal complexes with organic ligands are adsorbable at electrodes. This property can be utilized in adsorptive accumulation of metal chelates at an electrode, followed by the reduction of the adsorbed compound producing a peak on the voltammetric curve. This permits sensitive determinations of metal ions that cannot be determined by anodic stripping voltammetry or which are very difficult or even impossible to determine by conventional polarographic or voltammetric methods, e.g. Al, Be, Sr, Ca, Mg, Si, and B. The metal ion analyses are mostly applied to waters, mainly sea water. Competition between the added ligand and naturally occurring complexing material provides a means of evaluating the complexing ability of sea water [14]. In general, stripping methods are frequently used for metal speciation in waters. AdSV can sometimes be used to determine the number of cations [15] (such as Cu^{2+} in complexes), where the positive potential at which adsorption accumulation is carried out prevents the deposition of some ions (e.g. Pb^{2+}), that would interfere in anodic stripping voltammetry.The sensitivity of AdSV is often better than that of ASV, as the metal is not dissolved in mercury, but rather a monomolecular complex layer is formed on the electrode surface.

The method most extensively used in practice is the nickel determination at a mercury electrode, in the form of Ni-dimethylglyoximate. The first paper on this subject [16] was published in 1947 and describes the increase in the polarographic current at a dropping mercury electrode as a result of adsorption of the Ni-dimethylglyoximate. The AdSV determination of nickel can be carried out in various materials such as water, biological materials, foodstuffs, etc. [17], as well as in lipid fractions of biomaterials [18]. The determination limit in water is 1 $\mu g\ l^{-1}$ Ni^{2+}. It has been found useful in toxicological studies to determine nickel (and also lead, cadmium, and many others) in fingernails [19], as the concentrations are about one order of magnitude greater than in urine or blood. A glassy carbon electrode covered with a mercury film has also been used [20], e.g. for determinations of Ni ir biological materials, atmospheric dust in various regions, airborne ash and rain water. Cobalt can be determined with the same complex forming agent.

Uranium can be determined in water at concentrations from 0.5 μg l^{-1} tc 0.2 μg l^{-1} by employing a method based on the adsorptive accumulation of its pyrocatechol complex [21, 22]. Many papers are devoted to AdS\ determination of aluminium and beryllium, which cannot be determined b conventional polarography. Aluminium can be determined at concentration from 1 x 10^{-5} mol dm^{-3} to 1 x 10^{-7} mol dm^{-3} after binding it in the adsorbabl complex with alizarin violet N [23] or cupferron [24]. This method has bee applied to determination of aluminium in waters and to studies of aluminiu release from antacidant remedies in acidic solutions [25]. The latt measurements have shown that the daily intake of aluminium ions can reac values of 400 mg to 1 g with normally administered amounts of antaci containing aluminium. This quantity is higher by several orders of magnituc than the amount of aluminium released from aluminium cooking utensils (ingested from water and foodstuffs. For determination of beryllium in water: in a concentration range from 1 x 10^{-6} to 1 x 10^{-8} mol dm^{-3}, AdSV can be use after binding Be into the adsorbable complex with Beryllon II [26] c Beryllon III [27].

A detailed review on metal determination using AdSV has bee published in [28].

4.3.2.2 DETERMINATION OF ORGANIC COMPOUNDS

AdSV permits a relative simple study of organic compounds characterized by surface activity in a concentration range from 1 x 10^{-6} to 1 x 10^{-9} mol dm^{-3}. Nevertheless, even lower detection limits were obtained using mercury electrodes, e.g., for the pesticide DNOK (2-methyl-4,6-dinitrobenzol), 2.5 x 10^{-10} mol dm^{-3} at an accumulation time of 3 min [29]. Some examples of the AdSV determinations of environmentally important organic compounds are given in ***Table 1***.

This method can be employed in trace analysis of a great variety of surface active organic compounds. If the given compound contains an electrochemically reducible or oxidizable group, the peak current on the voltammetric curve, recorded after completion of the accumulation process, corresponds to the reduction (or oxidation) of the total quantity of the species accumulated at the electrode (and transported to the electrode during the potential scan).

Tab. 1: **Basic parameters for AdSV of some organic substances at a mercury electrode**

Compound	Supporting electrolyte	E_{acc}/V	E_p/V	Ref.
Nitrobenzene	B-R, pH 7	-0.20	-0.55	30
1,8-Dinitronaphthalene	B-R, pH 8	-0.20	-0.42	30
4,8-Dinitronaphthalene	B-R, pH 8	-0.20	-0.36 -0.46	30
2,4-Dinitro-1-naphthophenol	0.2M NaOH	-0.50	-0.72 -0.77	30
DNOK	B-R, pH 6.1	-0.20	-0.31 -0.44	29
Dinobuton	B-R, pH 6.1	-0.30	-0.46	29
Prometryne	B-R, pH 3.5	-0.70	-1.05	29
Ametryn	B-R, pH 3.5	-0.70	-1.02	29
Paraquat	ac, pH 4.6	-0.60	-1.13	31
Laurylsulphonate*	1M NaOH	-0.70	-1.20	15
Dodecylbenzenesulphonate*	1M NaOH	-0.70	-1.20	15
Oil products* and crude oil	1M NaOH	-0.70	-1.20	32

Only tensammetric adsorption/desorption peaks are obtained for electro-inactive compounds (see compounds labeled with the asterisk in ***Table 1***). The height of the tensammetric peak obtained is partly dependent on parameters similar to those governing electrolytic stripping voltammetry, and partly dependent on the surface-active properties of the particular compound. The stronger the adsorption, the higher and narrower are the peaks. The adsorption can be influenced by increasing the concentration of the supporting electrolyte, where salting-out effects of the compound also sometimes contribute to an increase in the capa-citive phenomena. The adsorptive stripping method can only be used when the compound yields a well developed tensammetric peak - without accumulation - at a concentration of about 10^{-5} - 10^{-6} mol dm^{-3}. Such compounds have usually an adsorption coefficient of about 10^3 mol^{-1} dm^3 or higher in the respective supporting electrolyte. At a DME, adsorption equilibrium is usually not established within the usual drop-time; when a SMDE is used with an accumulation period, equilibrium coverage of the electrode can be attained at concentrations lower than 10^{-5} mol dm^{-3}, bringing about a noticeable tensammetric effect. With weakly surface-active substances yielding tensammetric peaks at higher concentrations (ca. 10^{-1} or 10^{-2} mol dm^{-3}), the time necessary to attain an equilibrium coverage of the electrode is much shorter than the DME drop time, and thus application of an accumulation process is not effective.

As can be seen from the published papers, the range of compounds which can be determined by AdSV is very broad, involving pharmaceuticals, dyes, pesticides, biochemicals, DNA, detergents, crude or motor oils and emulsions. A number of substances that cannot be reduced polarographically can be determined after a derivatization procedure, introducing a reducible group, such as nitroso- , nitro-, etc.

4.3.3 Limitations of the ADsV method

One of the most serious complications in the use of AdSV can be the presence of other surface active compounds in the solution examined: competitive adsorption usually occurs, decreasing the height of the peak or suppressing the signal completely. These surfactants present in the sample can contribute to full coverage of the electrode surface when using long accumulation times. Interfering effects also depend on the nature of both the analyzed and interfering substance and on their concentration ratio. In the determination of trichlorobiphenyl [33] (tens of $\mu g\ l^{-1}$), a thousand-fold excess of Triton X 100 produced a 90% decrease of the signal. On the other hand, when the Triton X 100 concentration was comparable to that of trichlorobiphenyl, the signal was unaffected. The peak for 5×10^{-8} mol dm^{-3} nitrazepam, which is reduced at a mercury electrode, was not affected by the presence of 5 $\mu g\ ml^{-1}$ gelatine as a model-substance at $t_{acc} = 15$ s, but the peak decreased by 40 percent for $t_{acc} = 30$ s and completely disappeared for $t_{acc} = 60$ s. An increase in the gelatine concentration, similar to an increase in t_{acc}, leads to a decrease in the nitrazepam peak [30].

In general, the inhibitive effect of accompanying adsorbable substances can be suppressed by using short t_{acc} values. On the other hand, interferences can be removed, e.g., by using gel chromatography on Sephadex [30], ultrafiltration [34] or extraction methods. Extraction with diethyl ether is among the most popular separation techniques (mainly for isolation of organic compounds from body fluids). For more details, see [35, 36].

In determinations of metals bound in chelates, interferences from surface-active and other organic compounds can sometimes be prevented by prior destruction of them by irradiating the sample with UV light or boiling it with hydrogen peroxide. Only after this operation should the ligand solution be added to the sample. Serious complications in metal determination can also occur due to competitive adsorption of the ligand, the concentration of which must be in excess. Difficulties arise mainly if the peak potential of the ligand is very close to the peak potential of the metal chelate - the conditions are often not so ideal as in the above mentioned determination of nickel as the dimethylglyoximate.

4.4 Conclusions

The AdSV method leads to a great improvement in the sensitivity of polarography/voltammetry for the determination of surface-active organic compounds. As many organic substances possess such properties, AdSV has found extensive application. The method can be employed for concentrations of 1 - 200 μg l^{-1} (and sometimes from 0.1 μg l^{-1}). This sensitivity in determination of organic compounds is similar to that found for metal ions by the anodic stripping method and thus corresponds to a considerable extension of voltammetry in organic trace analysis. This mode of analysis is also extended to many metals which form adsorbable complexes. AdSV thus permits sensitive determination for metals that are difficult or impossible to determine by ASV. In general, ASV and AdSV are at present the most extensively used modes of polarographic/voltammetric analysis.

4.5 References

1. Kalvoda R.: Coll. Czech. Chem. Commun. *21*, 852 (1956).
2. Kalvoda R.: Chem. listy *54*, 1265 (1960).
3. Kalvoda R., Budnikov G. K.: Coll. Czech. Chem. Commun. *28*, 838 (1963).
4. Čermák V.: Coll. Czech. Chem. Commun. *24*, 831 (1958).
5. Kano K., Konse T., Nishimura N., Kubota T.: Bull. Chem. Soc. Japan *57*, 2383 (1984).
6. Koryta J.: Coll. Czech. Chem. Commun. *18*, 206 (1953).
7. Ikeda T., Ando S., Senda M.: Bull. Chem. Soc. Japan *54*, 2189 (1981).
8. Webber A., Shah M., Osteryoung J.: Anal. Chim. Acta *154*, 105 (1983).
9. Wang J., Peng T., Lin M. S.: Bioelectrochem. Bioenerg. *15*, 147 (1986).
10. Brown A.P.,Anson F. C.: Anal. Chem.: *49*, 1589 (1977).
11. Kovář M., Kalvoda R., Novotný L., Berka A.: Electroanalysis *5*, 171 (1993).
12. Wenrui Jin, Jixian Peng : J. Electroanal. Chem. *345*, 433 (1993).
13. Štulík K., Pacáková V.: in *Instrumentation in Analytical Chemistry*, Vol. 2. (Zýka J., Ed.). Ellis Horwood, London 1994. P. 13.
14. van den Berg C. M. G., Donat J. R.: Anal. Chim. Acta *257*, 281 (1992).
15. Kalvoda R.: Anal. Chim. Acta *138*, 11 (1982).
16. Komárek K.: Coll. Czech. Chem. Commun. *12*, 339 (1947).
17. Pihlar B., Valenta P., Nünberg H. W.: Z. Anal. Chem. *307*, 337 (1981).
18. Gemmez-Ľolos V., Neeb R.: Z. Anal. Chem. *327*, 547 (1987).
19. Gammelgaard B., Andersen J. R.: Analyst *110*, 1197 (1985).
20. Braun H., Metzger M.: Z. Anal. Chem. *318*, 321 (1984).
21. Lam N. K., Kalvoda R., Kopanica M.: Anal. Chim. Acta *154*, 79 (1983).
22. van den Berg C. M. G., Huang Z. Q.: Anal. Chim. Acta *164*, 209 (1984).

23. Stará V., Kopanica M.: Coll. Czech. Chem. Commun. *54*, 370 (1989).
24. Dostál A., Duong H. B., Kalvoda R.: Chem. Listy *86,* 847 (1992).
25. Kalvoda R., Duong H. B.: Čs. gastroenterologie *47,* 218 (1993).
26. Dostál A., Kalvoda R.: Chem. Listy *86,* 380 (1992).
27. Sum C. G., Wang J. Y., Hu W., Xie T. Y., Jin W. R.: Anal.Chim.Acta *259*, 319 (1992).
28. Paneli M. G., Voulgaropoulos A.: Electroanalysis *5*, 355 (1993).
29. Beňadiková H., Kalvoda R.: Anal. Lett. *17,* 1519 (1984).
30. Kalvoda R.: Anal. Chim. Acta *162,* 197 (1984).
31. Kalvoda R.: unpublished results
32. Kalvoda R., Novotný L.: Vodni hosp. (Czech) *11,* 291 (1984).
33. Lam N. K., Kopanica M.: Anal. Chim. Acta *161,* 315 (1984).
34. Tocksteinová Z., Kalvoda R.: Chem. Listy *82*, 1209 (1988).
35. Kalvoda R :in *Instrumentation in Analytical Chemistry*, Vol. 2. (Zýka J., Ed.). Ellis Horwood, London 1994.
36. Kalvoda R.: Fresenius J. Anal. Chem. 349, 565 (1994).

5
ELECTROCHEMICAL DETECTION FOR HIGH-PERFORMANCE SEPARATION TECHNIQUES AND FLOW ANALYSIS

Karel Štulík and Věra Pacáková

Abstract

The work carried out at the UNESCO Laboratory on electrochemical flow detection is briefly summarized, discussed and placed within a more general framework. Emphasis is placed on electrolytic (voltammetric and, to a lesser degree, potentiometric) detectors; conductometric detection is only discussed from the point of view of its application to reversed-phase HPLC of ionic compounds. Some theoretical aspects of electrochemical detection have been studied, as well as various detector designs, working electrode materials, their activation and (bio)chemical modification. Applications involve continuous-flow and flow-injection analyses and, primarily, HPLC determinations of many compounds of environmental and biological importance, such as some pesticides and their residues, organic and inorganic ions, amino acids and peptides, biogenic amines and related substances, pharmaceuticals and other substances. Future trends are outlined.

Key Words

flow detectors, voltammetry, potentiometry, conductometry, liquid chromatography, flow-injection analysis, monitoring, pollutants, biologically active compounds

5.1 Introduction

Analytical measurements in flowing liquids are steadily gaining importance in all fields, including environmental analysis, for several reasons [1-3]: Industrial and environmental control require rugged and reliable monitoring systems; continuous monitoring of substances in physiological fluids is often needed in biology and medicine; continuous-flow (CF) and flow-injection (FI) analysis [4] substantially facilitate automation of laboratory analyses and are less expensive and more versatile than laboratory robots; most analyses now require a high-performance separation step and high-performance liquid chromatography (HPLC) is among the most important methods in this respect (moreover, the electrophoretic techniques that are now experiencing an enormous development, have a lot in common with chromatography).

It should be emphasized that all the above methods of flow analysis (including HPLC) are based on the same hydrodynamic principles [1] and that an ideal detector should follow the concentration (mass) profile of the analyte(s) without any distortion. Therefore, the detection systems should satisfy several requirements [1, 5]: All the analyte interactions and signal handling procedures must be sufficiently rapid; the hydrodynamic conditions must be well defined and reproducible; the effective volumes must be comparable to the volume of a single theoretical plate in the system (ranging from millilitres for industrial analyzers to less than a nanolitre for capillary separations).

Of the many measuring methods available for flow detection, absorption spectrometry is by far most widely used and is replaced by other techniques only when its parameters, especially sensitivity and selectivity, are inadequate to the problem in hand. Electrochemical methods have always been and will be highly specialized approaches; however, when their applications are properly selected, they are unrivalled in the sensitivity, selectivity and a high precision and accuracy. The often cited problem of fouling of the sensors can be overcome by judicious choice of the measuring and activation conditions; moreover, in flow measurements (except for continuous monitoring) this problem is generally less serious, as the electrodes are continuously washed with the carrier liquid (mobile phase) and only occasionally exposed to zones of samples that may contain surface active components. On the other hand, interactions of the sensor with the test solution have made it possible to develop such important techniques like stripping voltammetry or the use of physically and (bio)chemically modified electrodes, tailored for specific purposes.

Of the various electrochemical methods, conductometry has found routine use in certain industrial monitors (e.g., for total electrolyte content) and in ion chromatography. Ion-selective electrodes are quite popular in FIA; their application in HPLC is hampered by their slow response. Voltammetry and coulometry are most generally applicable, because of their high sensitivity for certain classes of compounds and a high selectivity that can be regulated to a certain extent by variation in the experimental conditions. A systematic treatment of electrochemical flow detection, with a survey of applications up to the end of 1984, can be found in the monograph [1]; more recent advances have been reviewed, e.g., in [2, 3, 5].

5.2 Results and Discussion

A number of detector types has been designed, tested and applied in the UNESCO Laboratory, often in cooperation with workers of other organizations, both from the Czech Republic and from abroad. The results are briefly surveyed below.

5.2.1 Detector Designs and Their Properties

5.2.1.1 POTENTIOMETRY

Potentiometry is rarely used for chromatographic detection, but ion-selective electrodes (ISE's) and ion-selective field-effect transistors (ISFET's) are quite popular in CFA and FIA [1, 4, 5]. In our work, we have been primarily concerned with improving the measuring sensitivity, reproducibility and speed of analysis. For process control purposes, a simple thin-layer cell has been constructed with an ISE as the working electrode [6] and it has been demonstrated that sensitive and reproducible measurements are possible even if the industrial application requires that the cell volume be large, provided that the flow rate of the carrier liquid is sufficiently high. Large cell volumes are advantageous in continuous monitoring, as the signal approaches that obtained under steady-state conditions. The response time is decreased by making the hydrodynamic boundary layer thinner when using a higher flow rate which, however, must be optimized from the point of view of the hydrodynamic conditions in the whole flow system to attain a stable response. Typical values of the cell response rate are given in ***Table 1*** for an ISE with a fast response (a CN-ISE).

Tab. 1: **Time constants (T_k) and response volumes (V_{resp}) for potentiometric cells in dependence on the flow rate**

Flow rate (µl s^{-1})	Detection cell volume (µl)					
	1 500		100		25	
	T_k (s)	V_{resp} (µl)	T_k (s)	V_{resp} (µl)	T_k (s)	V_{resp} (µl)
1.5	135.0	200	75.0	110	75.0	110
3.0	75.0	225	47.5	150	47.5	140
6.0	40.0	240	25.0	150	25.0	150
12.0	21.6	260	15	180	15.0	180
22.0	12,5	275	7.5	165	6.0	130
47.0	9.0	420	5.0	235	2.5	120
92.0	5.0	460	4.0	360	2.0	180
180.0	4.0	720	2.5	450	1.5	270

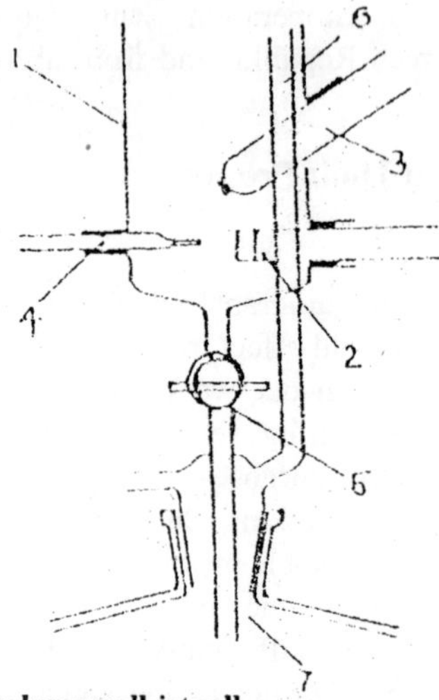

Fig. 1: **Large volume wall-jet cell**
1 - cylindrical glass vessel, 2 - working electrode, 3 - reference electrode, 4 - nozzle, 5 - solution outlet stopcock, 6 - tube maintaining constant solution level, 7 - waste bottle

A substantial improvement in the measuring sensitivity, reliability and sample throughput can be attained by employing a large-volume wall-jet cell. This type of cell was originally proposed for amperometric measurements [7, 8]; we designed a very simple version for potentiometric unsegmented flow analysis [9] (*Fig. 1*). The liquid jet issuing from the nozzle remains intact even at distances of the nozzle from the electrode surface of up to ca. 10 mm, while the effective working volume of the cell is small (less than the volume of the Prandtl layer). The stream of the test solution protects the electrode from the influence of the liquid in the cell and the horizontal position of the nozzle and the electrode prevents air bubbles from sticking to the electrode surface. This cell permits continuous regeneration of the electrode surface: e.g., an acidic solution of aluminium ions is placed in the cell to remove passivating layers from a fluoride ISE or a diaminocyclohexanetetraacetic acid (DCTA) solution is used for this purpose in measurements with a chloride ISE. When a sample is pumped through the nozzle, the electrode is shielded by the sample solution from the regeneration solution and the electrode responds to the sample; between samples, air is pumped through the nozzle, stirring the regeneration solution and hastening the cleaning process. Typical analytical parameters then involve limits of detection of 6.3×10^{-8} and 2.3×10^{-7} mol l^{-1} for fluoride and chloride respectively, with a RSD of less than 3 %, a sample throughput of 90 h^{-1} and the lifetime of the regeneration solution of at least 150 measurements.

Commercial chloride ISE's suffer from a high impedance and frequent surface passivation by interferents present in the sample solution (e.g., sulphur compounds, cyanide, bromide and iodide); the impedance problem can be alleviated by placing an impedance converter close to the ISE membrane [10]. Both the problems are readily coped with when using a chloride ISE prepared by chemically depositing a thin silver chloride film on the surface of an iodide ISE (the reaction with mercuric chloride in an oxidizing medium containing excess chloride), and placing it in the above large-volume wall-jet cell containing the DCTA cleaning solution [11]. In analyses of polluted ground waters, this system yielded very good results (a LOD value of 2.6×10^{-7} mol l^{-1}, an RSD from 0.6 to 3.0 %, 90 measurements per hour and a satisfactory correlation with an indirect AAS determination of chloride).

5.2.1.2 VOLTAMMETRY

Voltammetric cells most often operate amperometrically, i.e., a limiting current is measured at a constant potential, the extent of electrochemical conversion being low (from tenths to units of a per cent). Coulometric response at a constant potential is obtained either with large-area electrodes, such as porous electrodes, and sometimes also with microelectrodes and their arrays. The latter also permit the obtaining of three-dimensional recordings in the time and potential domains, usually by using rapid potential scans; the use of micro-electrode arrays as multichannel detectors is still in its infancy. For systematic discussion of all the aspects of voltammetric detection see refs. [1-3].

In designing a voltammetric detection system, three principal problems must be solved:

(a) A suitable electrode material must be found from the point of view of the measuring sensitivity, selectivity and reliability; the electrode materials may be physically or (bio)chemically modified to improve their performance (some procedure for activation of solid electrodes is always required).

(b) The cell geometry must primarily permit rapid response, a sufficiently small working volume and reproducible and convenient hydrodynamic conditions, both from the point of view of the electrochemical requirements and those connected with the analyte zone dispersion. The placement of the reference electrode is important from the point of view of the uncompensated ohmic drop which is usually large, as microcells exhibit high impedances. It should be emphasized that the requirements connected with the electrochemistry and zone dispersion are sometimes contradictory, especially in HPLC, and thus practical cell designs represent an optimal compromise.

(c) The measuring conditions must be found and optimized for particular applications.

We have dealt with all these aspects and a summary of our results follows.

Working Electrodes and Their Treatment

The basic electrode materials that we have used involve mercury for reductive detection and platinum, gold and, especially, various forms of carbon for oxidative detection. Whereas mercury usually does not require any pretreatment, all the solid electrodes must be suitably treated for sensitive and reproducible measurements. The activation procedures for solid electrodes, involving mechanical, heat, chemical and electrochemical approaches and their combinations, were recently reviewed [12].

In our measurements with various solid electrodes, we have mostly treated the electrode material by mechanical polishing (usually only with a newly prepared electrode and with the obvious exception of carbon fibres) and by the most common procedure of potential cycling within suitable limits and in a suitable, non-complexing solution. This procedure usually ends at a potential close to zero (*vs.* an SCE) and it is hoped that the surface is almost free of any functional groups. However, many electrode reactions are selectively enhanced by surface functionalities produced by oxidation of the electrode surface [3, 12] and so it is sometimes useful to preoxidize the working electrode; a certain drawback for flow measurements lies in the fact that this enhancement usually involves adsorption of the reactants at the electrode surface and thus the response rate is decreased. The increase in the measuring sensitivity is then not so pronounced as in batch measurements, but a considerable improvement in the selectivity (e.g., separation of dopamine and ascorbic acid signals, very important for *in vivo* measurements) can be fully utilized.

We have successfully employed laser irradiation [13] and application of short anodic current pulses [14] to produce modified carbon surfaces. The laser treatment is based on the heat evolved and leads to removal of a thin layer of adsorbed particles and the electrode material and a decrease in the amount of chemisorbed oxygen. The measuring sensitivity and selectivity are improved with many systems; however, the electrode roughness slowly increases on repeated irradiations and the electrode must periodically be repolished. On the other hand, the anodic treatment produces oxygen-containing functionalities on the carbon surface, primarily phenols, carbonyls and carboxyls that exhibit the above catalytic effects. With increasing charge passed, the concentration of surface phenols increases and that of carbonyls decreases and there is an optimal charge at which the electrode exhibits the highest activity; very high charges produce a layer of graphite oxide that no longer has redox properties, but behaves as an ion-selective membrane [15]. Both the methods can produce surfaces tailored for a given purpose, by finely tuning the irradiation or electrochemical conditions.

Carbon paste electrodes are quite popular, because they are very simply and cheaply prepared and are easily modified. However, for flow measurements they have several disadvantages, namely, poor mechanical strength, leaching of the diluent, especially in HPLC mobile phases containing organic solvents, and impossibility to polish them properly. We have obtained very good results by replacing liquid diluent by binders of polyvinyl chloride or a copolymer of chloroprene rubber and an alkylphenol resin and thus producing rugged composite electrodes [16]. These electrodes are easily polished and behave as a microelectrode array, thus producing an enhanced amperometric signal.

Another possibility of performing a highly selective detection is the use of selective chemical reactions of analytes with the ions produced by oxidation of a metal electrode. This principle has been known for decades in polarography where anodic waves are formed on oxidation of a dropping mercury electrode in the presence of substances forming complexes or poorly soluble compounds with the mercury ions. However, this system cannot be used for flow measurements, as the mercury ions are soluble and would be immediately transported from the electrode surface. On the other hand, when a copper electrode is polarized at slightly positive potentials in a solution whose pH is not lower than ca. 6.0, the electrode is not dissolved but a passivating bi-layer is formed on its surface, the inner layer consisting of highly insoluble cuprous oxide and the thicker, porous outer layer containing cupric ions in the form of various basic salts and cupric hydroxide. Cupric ions form stable complexes with many organic and some inorganic compounds and these ligands can then be determined in the same way, as is the classical polarographic principle. This approach has been used both in potentiometric and amperometric determinations (for the references see, e.g., the review [5]).

We studied this principle in detail [17] and applied it to selective and sensitive HPLC detection of amino acids [18], amines [17], short-chain peptides [19,20], some fodder biofactors [18] and ethylene thiourea in beverages [21]. The principle of this detection is schematically depicted in ***Fig. 2***. On an example of four amino acids, it is demonstrated in ***Table 2*** that sensitive detection requires not only a stable complex between the analyte and cupric ions, but primarily a rapid complex formation. The size of the analyte molecule also strongly affects the complexation; e.g., only short peptides (di- and tripeptides) can be detected. Peptides with protective terminal groups cannot be detected. This on one hand limits the applicability range, but on the other hand provides another dimension of selectivity.

A great number of detection systems involving chemically and biochemically modified electrodes have been described and applied with varying success (for recent surveys see, e.g., [1-3, 5]). We obtained good results in FIA determination of glucose and various phenols, using a thin-layer cell containing a Clark oxygen sensor whose membrane was modified with immobilized glucose oxidase and tyrosinase, respectively [22]. When comparing this system with that containing the enzymes immobilized in

Tab. 2: **Rates of complexation, complex stabilities, detection sensitivities and detection limits for detecting amino acids by amperometry with passivated copper electrode**

Test solution	A	A	A	A	B	C	D
Solute	histidine	serine	valine	aspartic acid	histidine	histidine	histidine
Reaction half-time at 25 °C (s)	0.4	1.5	1.9	2.3	no reaction	very slow	very slow
Rate constant ($l\ mol^{-1}s^{-1}$)	2.5×10^3	6.7×10^3	5.3×10^3	4.3×10^3	-	-	-
Stability constant							
log β_1	10.20	7.89	8.11	8.57	-	-	-
log β_2	18.10	14.48	15.35	15.35	-	-	-
Sensitivity ($nA\ ng^{-1}$)	4.6	2.5	1.5	1.4	-	0.86	0.70
Detection limit (ng in 20 μl)	0.4	0.5	4.1	13.0	-	2.5	3.0

Test solution compositions (A to C are based on 0.025M KH_2PO_4):
A: pH 6.8, 10% (v/v) methanol;
B: pH 3.0, 10% (v/v) methanol;
C: pH 6.8, 50% (v/v) methanol;
D: 0.025M K_3BO_3, pH 6.8, 10% (v/v) methanol

(a)

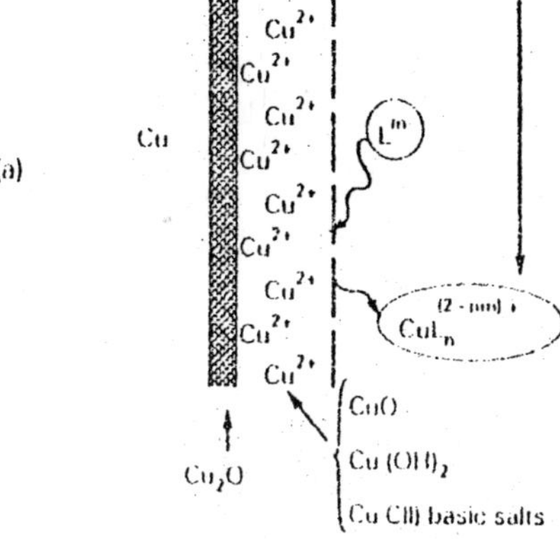

(b)

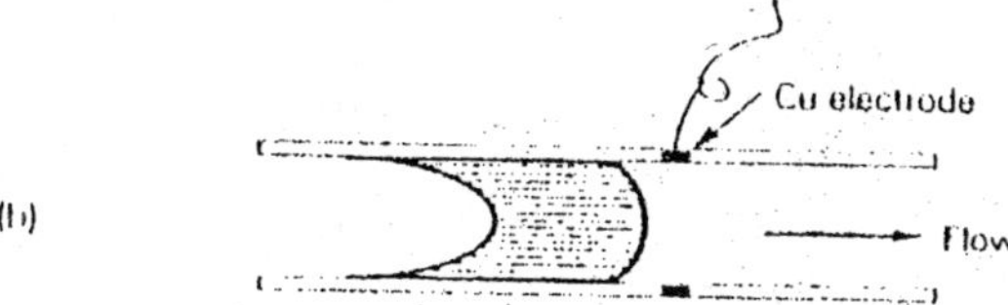

(c)

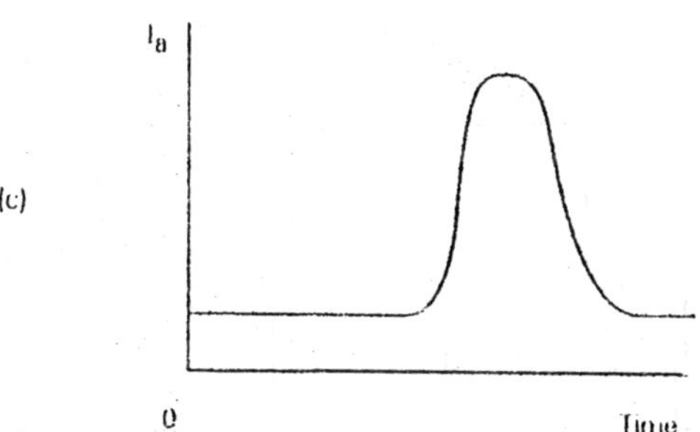

Fig. 2: **Detection with a passivated copper electrode on the basis of the complexation reaction**

$$Cu^{2+} + nL^{m-} \overset{\beta}{\longleftrightarrow} CuL_n^{(2-nm)+}$$

(a) - principle; (b) - sample profile; (c) - corresponding current signal obtained a preceding reactor, it was found that immobilization of the enzymes directly on the Clark sensor led to a higher sensitivity and a faster response, whereas the enzymes placed in the reactor permitted a greater sample capacity and a longer lifetime.

We also attempted to use in FIA measurements mercury film and gold film electrodes modified with tri-n-octylphosphine oxide (TOPO) which is a highly selective extraction agent for a number of ion associates; we applied them successfully in batch stripping determinations [23,24], but for flow measurements their response was too slow.

Cell Design

We have designed a number of cells with various geometries, tested them, compared their behaviour with the theoretical assumptions of hydrodynamic electrochemistry and applied them mostly to HPLC detection. Our original cell [25] was actually the first tubular system described in the literature whose volume (several microlitres) was suitable for FIA and HPLC; the later version [26] was improved in design and its volume was decreased to ca. 2 μl. Platinum working and counter electrodes were used in these cells. A further design [27] was of the thin-layer type which could be converted into the wall-jet configuration by turning the bottom part of the sandwich body by 180 degrees. Glassy carbon, carbon paste and carbon composite [16] electrodes were used and the working volume was further decreased to 0.7 μl. For reductive detection, a large cell was employed, feeding the analyte to a static mercury drop electrode through a horizontal jet placed in an immediate vicinity of the drop [28]. The effective volume of the cell was around 15 μl.

The best results have, however, been obtained with the cell employing a strand of carbon fibres as the working electrode [29]. Its perfected version [18] is depicted in ***Fig. 3***. The cell volume is around 0.1 μl. Alternatively, a tubular semimicroelectrode can be used instead of the strand of fibres (e.g., a copper electrode, see the previous section). The cell is directly attached to the outlet of a microelectrode array, i.e. the signal dependence on the mobile phase flow rate is suppressed, the response rate is very high (time constant is around 0.1 s) and the signal is enhanced compared with a normal-size electrode. In addition, the effects of electrode poisoning are almost non-existent; e.g., the detector worked more than one year without any deterioration in the signal magnitude and reproducibility, with only a few-minute stabilization in the beginning of each day.

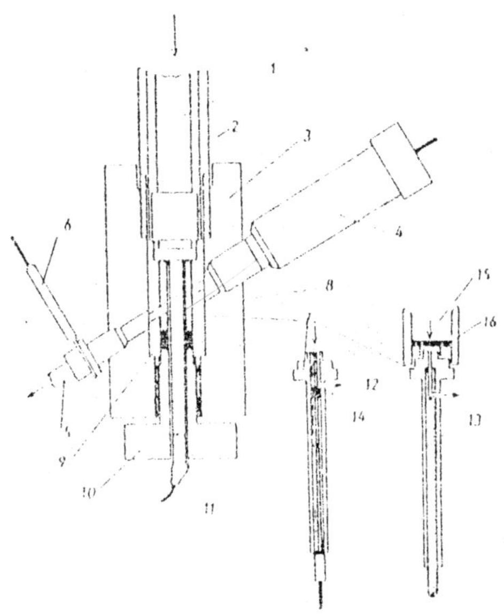

Fig. 3: **Schematic representation of cell with carbon fibre array working electrode (cell connected directly to the column)**

1 - glass column, 2 - metal mantle, 3 - detector plastic body, 4 - reference electrode, 5 - stainless steel outlet tube serving as the counter electrode, 6 - lead to the counter electrode, 7 - PTFE tube containing the working electrode (fibres), 8 - porous ceramic tube, 9 - silicone rubber seal, 10 - plastic screw closing the working space, 11 - lead to the working electrode, 12 - fibres, 13 - rod of the electrode material, 14 - silicone rubber seal, 15 - column packing material, 16 - metallic frit

The performance of the above detectors was compared [29, 30]. ***Table 3*** surveys important operational characteristics and ***Fig. 4*** demonstrates that the detector response does not differ much from that calculated from the appropriate equations for the limiting current, except for the tubular system with a platinum electrode and a thin-layer detector containing a carbon paste electrode; these deviations can be attributed to passivation of the electrodes. The advantages of the carbon fibre microelectrode array over the other detectors are obvious, while there is no pronounced difference in the performance of all other detector types.

Tab. 3: **Operational characteristics of various detection cells**

Cell type/ working electrode	Detection limit (ng)	Linear dynamic range (ng)	Calibration dependence		Time constant T_k (s)	Response volume (μl)	Geometric volume (μl)
			slope (nA/ng)	corr. coeff.			
Polarography /HMDE	15 [a]	100-30 000	0.99	1.000	2.2 [b]	36.7	8
Polarogr./conical HMDE	3 [a]	10-30000	0.53	0.999	2.3 [b]	38.3	8
Polarogr./conical HMDE[c])	3 [a]	10-25 000	0.27	1.000	0.6 [b]	10.0	8
Polarography/HMDE, large wall-jet cell	6 [d]	-	6.2	0.997	33.7 [e]	18.5	15
Tubular/Pt	0.3 [f]	0.3-2 500	1.3	0.998	1.7 [e]	8.5	2.3
Thin-layer/glassy C	0.5-1.0 [f]	1.0-5 000	2.6	0.999	0.8 [e]	3.8	0.7
Thin-layer/C paste	0.5-1.0 [f]	1.0-5 000	1.2	0.997	1.7 [e]	8.5	0.7
Wall-jet /glassy C	0.3 [f]	0.3-3 000	2.3	0.998	1.0 [e]	5.0	0.4
Wall-jet/C paste (nujol)	0.03 [f]	0.03-3 000	2.3	0.958	0.9 [e]	4.3	0.4
Wall-jet/comp. C	0.1 [f]	0.1-3 000	3.0	0.999	-	-	0.4
Carbon fibres/glassy C	0.003 [g]	0.003-50	3.7	0.997	0.2 [e]	1.0	0.4

Notes: a - nitrobenzene, b - flow rate 1 ml min^{-1}, c - wall-jet system, d - picric acid, e - flow rate 0.3 ml min^{-1}, f - ephinephrine, g - benzidine. Amperometric measurement at a constant potential corresponding to the limiting current.

For work with open-tubular capillary columns in HPLC, all the above cells have too large volumes. For this purpose, a micro-wall-jet cell has been constructed [31] (***Fig. 5***) and its applicability demonstrated.

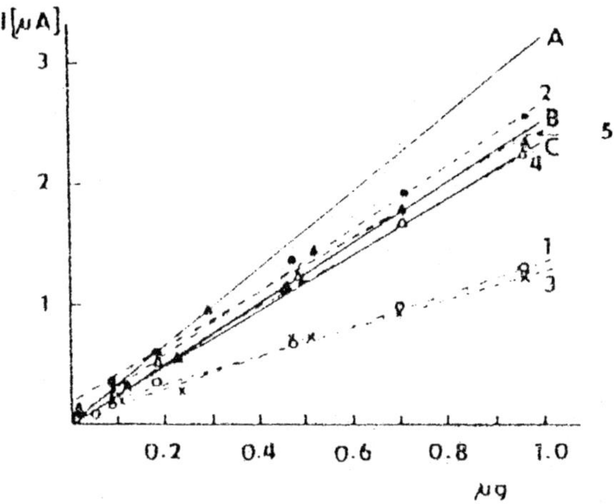

Fig. 4: **Experimental and theoretical calibration dependences**

A,B,C - theoretical plots for the thin-layer, wall-jet and tubular system, respectively. Experimental curves: 1 - tubular (Pt), 2 - thin-layer (glassy carbon), 3 - thin-layer (carbon paste), 4 - wall-jet (glassy carbon), 5 - wall-jet (carbon paste)

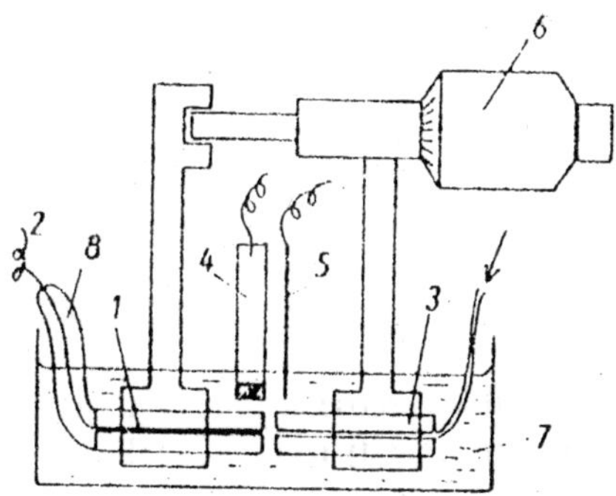

Fig. 5: **Micro-wall-jet detector for capillary columns**

1 - glass capillary with carbon composite working electrode, 2 - platinum lead to the working electrode, 3 - glass capillary containing the end of a fused silica capillary column fixed with epoxy resin, 4 - reference electrode, 5 - platinum counter electrode, 6 - micrometer screw, 7 - vessel with base electrolyte, 8 - chloroprene rubber insulation

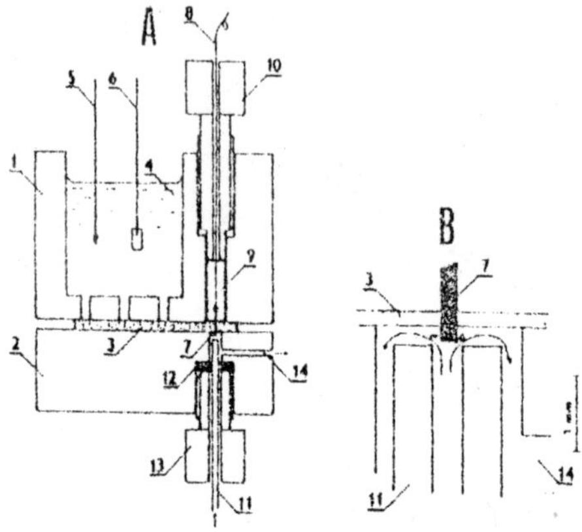

Fig. 6: Amperometric cell with a solid polymer electrolyte (spe) membrane
(A) overall view; (B) expanded view of the detection space
1, 2 - two parts of the electrode body, 3 - spe membrane, 4 - auxiliary electrolyte solution, 5, 6 - reference and auxiliary electrodes, 7 - working wire electrode, 8 - electrical contact to the working electrode, 9 - methacrylate cylinder, 10 - fixing screw, 11 - inlet capillary, 12 - seal, 13 - fixing screw, 14 - outlet

The detector versatility is greatly extended by using a dual working electrode arrangement (see, e.g., [1]). For this purpose, a thin-layer cell has been constructed containing a split-disk glassy carbon working electrode [32]. The two halves of the disk are separated by a thin insulating layer and act as two independent working electrodes that can be polarized at different potentials and/or differently pretreated. The disk can be turned, so that either a parallel or a series arrangement is obtained, permitting, e.g., background current subtraction, generation of a derivative with better electrochemical properties at the upstream electrode, differential recordings, etc.

Voltammetric detection is somewhat limited by the necessity of working in electrically conductive solutions; this problem is partially alleviated by using microelectrodes. Another approach uses cells in which the working, counter and reference electrodes are not connected by the test solution, but by a solid polymer electrolyte. Our cell with a platinum wire semimicroelectrode and Nafion (DuPont) as the solid polymer electrolyte [33] is depicted in ***Fig. 6***. Its response is reasonably fast (a time constant of 1.5 s at a flow rate of 0.7 ml min^{-1}), with a good linearity, a sufficient linear dynamic range of at least three concentration decades, a good sensitivity with LOD values in subnanogram range for common LC sample volumes and a good precision (RSD values of 1 to 5 %). In this way, the applicability of amperometric detection is substantially widened, e.g., to normal-phase HPLC and methods with programmed composition of the mobile phase.

Measuring Conditions

In addition to simple amperometry or coulometry, various waveforms have been applied to working electrodes, usually to improve the sensitivity and/or selectivity by pulse polarization, or to obtain three-dimensional, current-potential-time recordings (for a survey, see, e.g., [1-3, 5]. A simple square-wave polarization of a low frequency (below 1 Hz) and current sampling can be used for continuous reactivation of the electrode surface [25] (***Fig. 7***).

A dual electrode system can be used for a two-channel detection [32] (see the previous section). A similar effect can be attained by simultaneous use of several detectors. An example of the use of a series arrangement of a UV photometric, voltammetric and polarographic detector [34] is given in ***Fig. 8***.

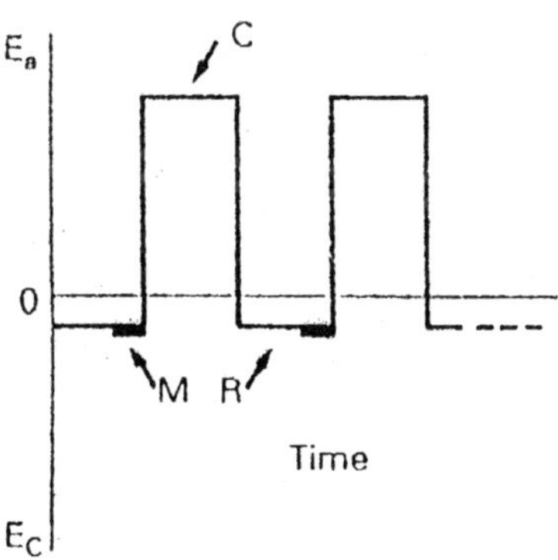

Fig. 7: Electrode polarization by alternate measuring and cleaning pulses
M - measuring interval, C - cleaning interval, R - relaxation interval

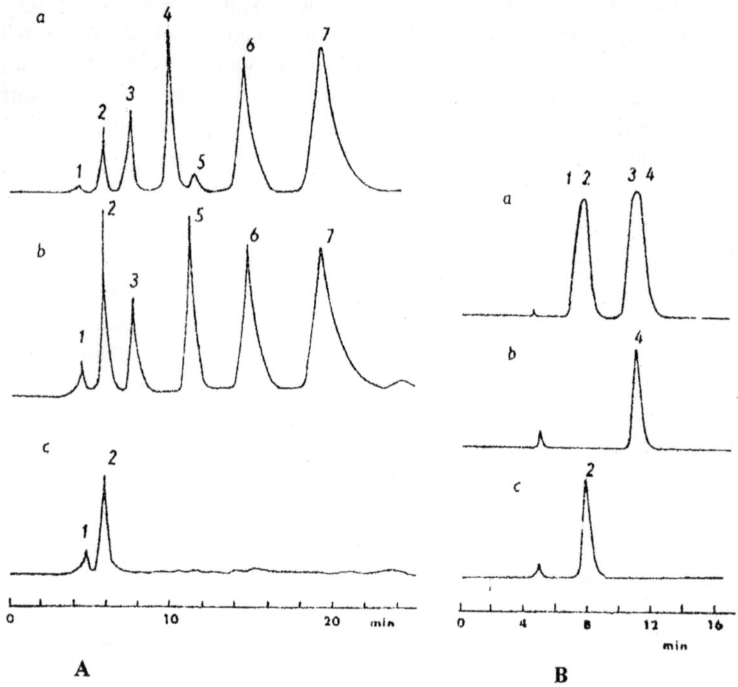

Fig. 8: **(A) Separation of several pyrimidine derivatives**

a - UV photometric detection at 254 nm; b - voltammetric detection, +1.4 V (*vs.* Ag/AgCl); c - voltammetric detection, 0.8 V (*vs.* Ag/AgCl);

1 - hold-up time; 2 - 2,4,5-triamino-6-hydroxypyrimidine; 3 - 2-amino-4,6-dihydroxypyrimidine; 4 - uracil; 5 - 4,5,6-triaminopyrimidine; 6 - 6-amino-4-hydroxy-2-mercaptopyrimidine; 7 - mercaptouracil

(B) Chromatogram of cytosine(1), 6-azacytosine (2), uracil (3) and 6-aminouracil (4)

a - UV photometric detection at 254 nm; b - voltammetric detection, +1.4 V (*vs.* Ag/AgCl); c - polarographic detection at 1.0 V (*vs.* Ag/AgCl)

Surface-active substances that are electroinactive (e.g., phospholipids) can be detected tensametrically at a mercury electrode [35]. However, the method is rather empirical and sensitive to interferences.

5.2.2 Applications

The principles and instrumentation developed, as described in the previous sections, have been applied to many substances and sample types. The predominating method has been HPLC, in which electrochemical detection has always been used in series with UV photometric detection; this substantially extends the power of identification and quantitation of analytes, as two independent signals are available. Some works describe FIA analyses, either potentiometric or amperometric with enzyme modified electrodes or with a preceding enzyme reactor. A brief survey is given below.

(a) Substances of biological and clinical importance

An obvious application of oxidative amperometric detection is HPLC of various neurotransmitters, primarily catecholamines and their metabolites. We studied and applied this technique a number of times and a survey can be found in refs. [1, 36]. Serum analyses have been used, e.g., in obesity studies. Analyses for amino acids and short-chain peptides were used mostly in the checking of the purity of preparations for various purposes [19, 20, 37] and also applied to the study of radiolysis of tryptophan [38]. Separation and detection of various pyrimidine derivatives were studied for purposes of biological research [34]. Hallucinogenic indoleamines were determined in the cerebrospinal fluid and in mushrooms, for medical purposes [39]. Equine estrogens were determined by HPLC with amperometric detection [53]. In the pharmaceutical field, HPLC methods have been developed and applied to analysis of tricyclic neuroleptics [40], thiobenzamide derivatives [41] and narciclasine [42]. A simple HPLC separation, followed by reaction in a reactor containing alcohol oxidase and amperometric detection of hydrogen peroxide produced, was used for a simple, selective, fast and sensitive determination of ethanol in blood [43]. Flow-injection analysis has been used for determination of phospholipids with tensammetric detection [35].

(b) Substances important in agriculture, toxicology and hygiene and in environmental protection

Fodder biofactors were determined by HPLC with amperometric detection on a copper electrode [18]. In analyses for pesticides and their residues, ethylene thiourea (a degradation product of carbamate pesticides) was determined by HPLC with amperometric detection on a copper electrode [21], while s-triazines and their degradation products were detected at a carbon fibre array [44]. HPLC analyses were applied to determination of carcinogenic aromatic amines [45] and azobenzenes [46] and their degradation products, and further to N-nitroso-N-methylaniline derivatives [47]. Environmentally important phenols were determined by HPLC [48] and by a selective FIA procedure with detection at a biochemically modified Clark sensor [22]. With respect to analyses of environmental samples, mainly natural waters, various HPLC methods have been developed and tested, both for inorganic anions [49, 50] and inorganic cations [51,52]. Combinations of UV/VIS photometric, conductometric and amperometric detection have been used. Flow-injection analysis with ISE potentiometric detection was used for determination of fluoride and chloride in contaminated natural waters [9, 11].

5.3 Conclusions

The field of electrochemical detection is now well established. It should always be borne in mind that electrochemical techniques are specialized approaches whose range of applications cannot compete with spectral methods. However, for certain groups of compounds and for certain types of samples they are unrivalled and their inherent sensitivity and tunable selectivity can be used to full advantage. Especially powerful are combinations of electrochemical measurements with high-performance separations; a great potential now lies in rapidly developing capillary electrophoresis.

The most important trends for future development of electrochemical flow detection involve the following:

(a) Use of micro- and ultramicroelectrodes and their arrays.

(b) Physico-chemical and biochemical modification of electrodes and the use of selective chemical reactions between analytes and electrodes.

(c) Rapid potential-scan techniques, dual and multichannel detection.

(d) Combinations of several detection methods, spectroelectrochemistry and spectroelectrochemical derivatization.

(e) Use of membrane electrodes, solid-polymer electrolyte cells and microelectronic sensors.

(f) Charge-transfer between two immiscible electrolyte solutions.

5.4 References

1. Štulík K., Pacáková V.: *Electroanalytical Measurements in Flowing Liquids*, E. Horwood, Chichester 1987.
2. Štulík K: Anal. Chim. Acta *273*, 435 (1993).
3. Štulík K.: *Some Ways of Improving the Sensitivity and Selectivity of Flow Electroanalysis*. In: *Reviews on Analytical Chemistry, Euroanalysis VIII* (Littlejohn D., Thorburn D., Eds.). Royal Society of Chemistry, Cambridge 1994.
4. Růžička J., Hansen E. H.: *Flow Injection Analysis*. J. Wiley, New York, 2nd ed., 1988.
5. Štulík K., Pacáková V.: Selective Electrode Revs. *14*, 87 (1992).
6. Peták P., Štulík K.: Anal. Chim. Acta *185*, 171 (1986).
7. Horvai G., Tóth K., Fekete J., Pungor E.: Euroanalysis IV, Helsinki 1981.
8. Gunashingham H., Fleet B.: Anal. Chem. *55*, 1409 (1983).
9. Lexa J., Štulík K.: Talanta *38*, 1393 (1991).
10. Langmaier J., Štulík K., Kalvoda R.: Anal. Chim. Acta *148*, 19 (1983).
11. Lexa J., Štulík K.: Talanta *41*, 301 (1994).
12. Štulík K.: Electroanalysis *4*, 829 (1992).
13. Štulík K., Brabcová D., Kavan L.: J. Electroanal. Chem. *250*, 173 (1988).
14. Mattusch J., Hallmeier K.-H., Štulík K., Pacáková V.: Electroanalysis *1*, 405 (1989).
15. Majer V., Veselý J., Štulík K.: J. Electroanal. Chem. *45*, 113 (1973).
16. Štulík K., Pacáková V., Stárková B.: J. Chromatogr. *213*, 41 (1981).
17. Štulík K., Pacáková V., Kang Le, Hennissen B.: Talanta *35*,455 (1988).
18. Štulík K., Pacáková V., Weingart M., Podolák M.: J. Chromatogr. *367*, 311 (1986).
19. Štulík K., Pacáková V., Jokuszies G.: J. Chromatogr. *436*, 334 (1988).
20. Heping Wang, Pacáková V., Štulík K.: J. Chromatogr *509*, 245 (1990).
21. Heping Wang, Pacáková V., Štulík K.: J. Chromatogr. *457*, 398 (1988).
22. Pacáková V., Štulík K., Brabcová D., Barthová J.: Anal. Chim. Acta *159*, 71 (1984).
23. Lexa J., Štulík K.: Talanta *32*, 1027 (1985).
24. Lexa J., Štulík K.: Talanta *36*, 843 (1989).
25. Štulík K., Hora V.: J. Electroanal. Chem. *70*, 253 (1976).
26. Štulík K., Pacáková V.: J. Chromatogr. *192*, 135 (1980).
27. Podolák M., Štulík K., Pacáková V.: Chem. Listy *76*, 1106 (1982).
28. Štulík K., Pacáková V., Podolák M.: J. Chromatogr. *262*, 85 (1983).
29. Štulík K., Pacáková V., Podolák M.: J. Chromatogr. *298*, 225 (1984).
30. Štulík K., Pacáková V.: J. Chromatogr. *208*, 269 (1981).
31: Hetem M. J. J., Claessens H. A., Leclercq P. A., Cramers C. A., Pacáková V., Štulík K.: Proc. 8th Int. Symp. on Capillary Chromatography, Vol. II (Sandra P., Ed.). Hüthig, Heidelberg 1987. P. 1122.
32. Mattusch J., Werner G., Štulík K., Pacáková V.: Electroanalysis *2*, 443 (1990).
33. Loub L., Opekar F., Pacáková V., Štulík K.: Electroanalysis *4*, 447 (1992).
34. Štulík K., Pacáková V.: J. Chromatogr. *273*, 77 (1983).
35. Emons H., Schmidt T., Štulík K.: Analyst (London) *114*, 1593 (129(89).
36. Štulík K., Pacáková V.: *Electrochemical Detection of Catecholamines and Related Compouds*. In: *Quantitative Analysis of Catecholamines and Related Compounds* (Krstulović A. M., Ed.). E. Horwood, Chichester 1986.

37. Štulík K., Pacáková V., Heping Wang: J. Chromatogr. *552*, 439 (1991).
38. Štulík K., Pacáková V., Weingart M., Vlasáková V.: J. Chromatogr. *354*, 449 (1986).
39. Kysilka R., Wurst M., Pacáková V., Štulík K., Haškovec L.: J. Chromatogr. *320*, 414 (1985).
40. Pacáková V., Štulík K., Tomková H.: J. Chromatogr. *298*, 309 (1984).
41. Říhová M., Pacáková V., Štulík K., Waisser K.: J. Chromatogr. *361*, 347 (1986).
42. Švagrová I., Štulík K., Pacáková V., Caliceti P., Veronese F. M.: J. Chromatogr. *563*, 95 (1991).
43. Pacáková V., Štulík K., Kang Le, Hladík J.: Anal. Chim. Acta *257*, 73 (1992).
44. Pacáková V., Štulík K., Příhoda M.: J. Chromatogr. *442*, 147 (1988).
45. Barek J., Pacáková V., Štulík K., Zima J.: Talanta *32*, 279 (1985).
46. Burcinová A., Štulík K., Pacáková V.: J. Chromatogr. *389*, 397 (1987).
47. Barek J., Pham Tuan Hai, Pacáková V., Štulík K., Švagrová I., Zima J.: Fresenius J. Anal. Chem. *350*, 678 (1994).
48. Tesařová E., Pacáková V., Štulík K.: Chromatographia *23*, 102 (1987).
49. Pacáková V., Štulík K., Mingjia Wu: J. Chromatogr. *520*, 349 (1990).
50. Mingjia Wu, Pacáková V., Štulík K., Sacchetto G. A.: J. Chromatogr. *439*, 363 (1988).
51. Janoš P., Štulík K., Pacáková V.: Talanta *38*, 1445 (1991).
52. Janoš P., Štulík K., Pacáková V.: Talanta *39*, 29 (1992).
53. Novaković J., Tvrzická E., Pacáková V.: J. Chromatogr. *678*, 359 (1994).

6

ELECTROCHEMICAL GAS SENSORS

František Opekar

Abstract

Gas sensors with metallized membrane electrodes have been employed for determination of various sulphur-containing gases and for compounds in solution (**CN^-**, **SO_3^{2-}**, **NO_3^-**) which can be transferred into the gas phase in the form of an electroactive volatile product through a chemical reaction. Construction and characteristics of **H_2** and **NO_2** solid-state sensors with a solid polymer electrolyte (Nafion) are also discussed. Laboratory sources of **SO_2**, HCN, CO** and **H_2**, based both on permeation of gas from aqueous solution and electrolytic generation and used for calibration of gas sensors, are described.

Key Words

gas sensor, metallized membrane, solid polymer electrolyte, gas source

6.1 Introduction

Chemical sensors of gaseous compounds are important components of many environmental information systems. Gas sensors can be based on various physical, physico- and biochemical principles [1]. An important group of gas sensors is based on electrochemistry; Severinghaus-type potentiometric [2] and Clark-type amperometric [3] sensors are typical examples of gas sensors containing liquid electrolyte. Solid-state gas sensors containing no liquid component [4] are important from practical point of view, as they belong among the most robust chemical sensors.

Electrochemical sensors often suffer from instability of the response signal during long-term measurements. On the other side, because of their relatively high sensitivity, simple operation, low power consumption and low price, they have found use primarily in the construction of personal hand-held gas detectors and dosimeters. Therefore, new sensitive electrode materials, sensor designs and working conditions improving gas sensor characteristics are continuously sought and studied. The periodically appearing reviews dealing with all aspects of chemical sensors [5] clearly document constant interest in the field of development, testing and application of electrochemical gas sensors.

Our research in the field of electrochemical gas sensors is directed mainly to the three following areas:

a) gas sensors with a metallized membrane electrode;
b) solid-state sensors based on a solid polymer electrolyte;
c) gas sources for calibration of gas sensors.

6.2 Analytical Applications of Metallized Membrane Electrodes

As with other types of membrane electrodes, the main advantages of metallized membrane electrodes in electroanalytical chemistry are found in (1) the determination of gaseous substances in gaseous media and liquid media where the gaseous substances are either present or can be suitably generated from the substances present, and (2) the effective elimination of contact between the indicator electrode and substances in the analyzed medium that would decrease its activity or interfere in the determination. Analytical applications of metallized membrane electrodes have been summarized in the reviews [6, 7].

The metallized membrane electrodes used in our studies consisted of a porous metal layer formed on the surface of a porous Teflon membrane. The metallized surface of the membrane was immersed in an electrolyte solution and the other side was in contact with the analyzed medium (gas or liquid). A principal scheme of the detection principle is given in ***Fig.1***. Gaseous compound (A) passes from the analyzed medium through the membrane pores to the metallized side where it is oxidized or reduced and the corresponding current is measured (amperometric measurements), or it reacts with a component of the electrolyte to change its equilibrium concentration at the electrode surface; this change appears as a change in the potential measured (potentiometric measurements).

We have used metallized porous membrane electrodes (MePME) for determination of many compounds in various gaseous or liquid samples.

6.2.1 MePME Applications

Sulphur-containing gases [8]: An equivalent amount of H_2S formed by non-catalytic reductive pyrolysis is determined by electrochemical oxidation at an AuPME. The method has been tested both for the determination of total sulphur content and individual gases after separation on a chromatographic column and postcolumn pyrolysis (sensitivity 1.12 μA/ppm S, detection limit 4 ppb).

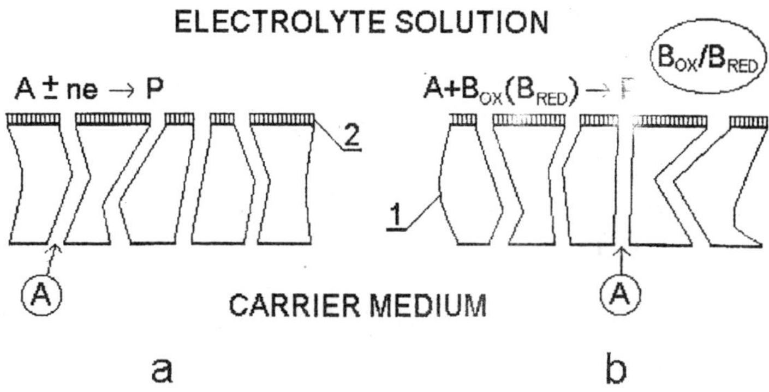

Fig. 1: **The principle of amperometric (a) and potentiometric (b) detection at MePME**

1 - porous membrane, 2 - metal layer, A - gas molecule, B - active electrolyte component

Total sulphur and nitrogen in crude oil materials [9]: Products of oxidative pyrolysis are detected amperometrically at an AuPME. In the determination of sulphur, the gaseous products are fed directly to the electrode; in the determination of nitrogen, the gaseous products are passed through a column containing PbO_2 in which SO_2 is removed and NO is converted into NO_2 (sensitivity 8.4 nA/ppm S and 4.2 nA/ppm N, detection limit 2.5 ppm S and 7.8 ppm N, analysis time 10-15 s).

Cyanide ions [10]: Samples are injected into a reaction solution containing an acid. The HCN liberated is transferred into the carrier gas (nitrogen) passing through the reaction solution and fed to an AgPME immersed in a buferred solution (pH = 9.5) of $KAg(CN)_2$. Some HCN passes through the membrane and affects the equilibrium concentration of Ag^+ ions in the vicinity of the AgPME surface. The corresponding change in the electrode potential is related to the amount of cyanide in the sample (sensitivity 1.24 mV/ng CN^-, detection limit 0.15 ppb CN^-; the detection limit and the sensitivity depend on the $KAg(CN)_2$ concentration). This method of analysis was named pneumatopotentiometry; for a scheme of the apparatus see ***Fig. 2***.

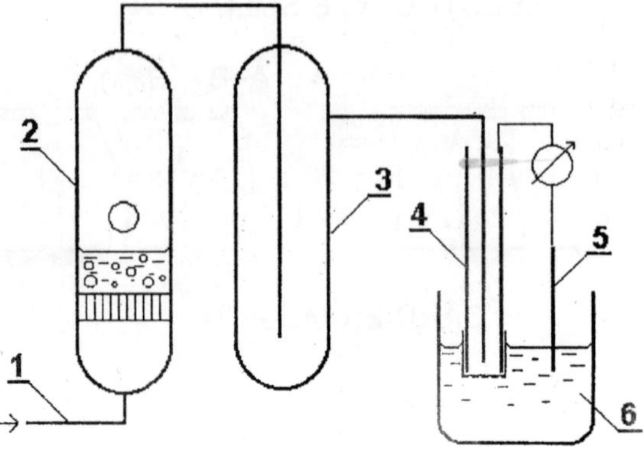

Fig. 2: **Scheme of the apparatus for pneumatopotentiometric analysis**
1 - inlet of carrier gas, 2 - reaction vessel equipped with a rubber septum for sample introduction containing reagent solution, 3 - separating vessel, 4 - MePME, 5 - reference electrode, 6 - electrolyte solution

Sulphur dioxide and sulphites [11]: Sulphur dioxide absorbed in a solution of sodium tetrachloromercurate or sodium hydroxide has been determined pneumatopotentiometrically. Sulphur dioxide is released from the absorption solution by the reaction with an acid and is stripped with nitrogen and led to an AuPME immersed in a solution of iodine monochloride. The SO_2 penetrating through the membrane pores reduces ICl_2 to I_2. The redox potential governed by the ratio of ICl_2^-/I_2 is sensed by the AuPME and its change is a measure of the SO_2 concentration in the absorption solution (sensitivity 9 mV/ng SO_2, detection limit 0.02 ppb; these values holds for a 10^{-5}M ICl_2 solution).

Sulphur dioxide in the atmosphere [12]: Sulphur dioxide can be preconcentrated by absorption from litres of air into microlitres of an absorption solution (sodium tetrachloromercurate), using a special enrichment unit [13] operating continuously on the basis of gas extraction into a polydisperse aerosol. The concentration of sulphur dioxide in the absorption solution is affected by the adjustable ratio of air and solution volume flow rates; a ratio of 10^4 was used. Sulphur dioxide in the absorption solution can be determined by pneumatopotentiometry (see above) and also pneumatoamperometrically, using the same apparatus and working procedure as that employed for pneumatopotentiometry. The SO_2 released from the absorption solution by acidification is transferred into the carrier gas and detected at an AuPME immersed in a 0.1M K_2SO_4 solution by electrochemical oxidation at 0.6 V *vs.* SCE. The oxidation current can be related to the SO_2 concentration in the air (detection limit 0.87 $\mu g\ m^{-3}$).

Pneumatoamperometric analysis can be modified so that volatile gaseous products are not transferred from the reaction solution into the carrier gas, but the reaction solution itself acts as the carrier medium. The sample is injected into the reactant stream which flows over the unmetallized side of an MePME. The volatile product formed is transferred from the reaction solution to the gas phase which fills the membrane pores. The slowest step in pneumatoamperometry with carrier gas - the transfer of the volatile product to the gas phase - is thus eliminated, resulting in a higher sensitivity and an increased sampling rate.

Nitrates [14]: The approach just mentioned has been used for a flow-through amperometric detector based on the electrochemical reduction of nitrates at a silver cathode. The nitric oxide formed is detected by a downstream AuPME. Nitrates were mostly determined in samples containing 5×10^{-2}M H_2SO_4 (sensitivity 0.85 nA/μmol l^{-1}, limit of detection 2.7×10^{-7}M NO_3^-). This method has been modified to eliminate the effect of nitrite (principal interferent) and permit its simultaneous determination in the presence of nitrate.

6.3 Solid-state Sensors

Electrochemical sensors with solid electrolytes permit the construction of miniature, rugged detectors that cannot be damaged by leaking of liquid electrolytes and corrosion. Solid polymer electrolytes are often employed for construction of solid-state gas sensors which operate under normal, laboratory temperature. One of the most common solid polymer electrolytes is the poly(tetrafluoroethylene)-poly(sulfonylfluoride vinylether) copolymer, well known under the trade name Nafion (DuPont).

6.3.1 Detection of gases

A Nafion membrane coated with chemically deposited platinum on both sides has been used for construction of an amperometric hydrogen sensor [15]. One side of the membrane (see ***Fig. 3***) is exposed to the test gas containing a low concentration of hydrogen in air (the indicator electrode), the other side is in contact with pure air thus forming the counter (or reference) electrode of the two-electrode galvanic cell

Pt (H_2, air) | NAFION | Pt (air)

Under the open-circuit conditions, the indicator electrode attains a mixed potential which is determined by the rates of the diffusion-controlled oxidation of hydrogen and the kinetically-controlled reduction of oxygen or Pt surface oxides, whereas only the two last reactions control the counter electrode potential. As a result of the different electrode potentials, an electric current flows in the short-circuited cell and is proportional to the hydrogen concentration. The sensor sensitivity is 8 nA/ppm H_2, detection limit is 30 ppm.

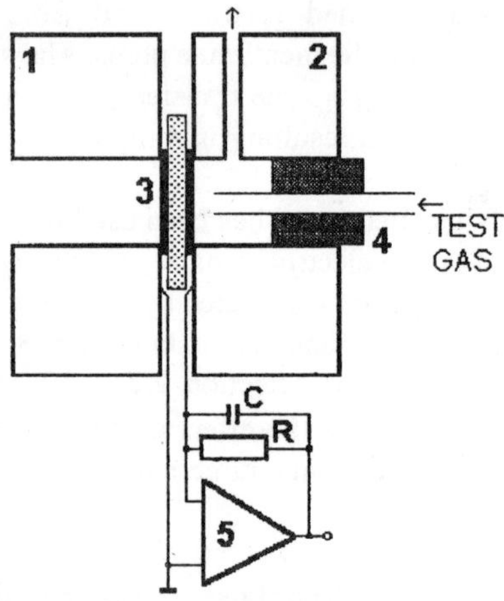

Fig. 3: **Scheme of the hydrogen solid-state sensor**

1, 2 - two parts of the sensor body, 3 - metallized Nafion membrane, 4 - stopper with the inlet tube, 5 - current meter

A certain drawback of this sensor is the necessity of using a reference gas (air that does not contain hydrogen) with the counter electrode. Therefore, we investigated the factors that influence the open-circuit potential of a Pt/Nafion electrode in air containing a low concentration of hydrogen [16]. The Pt electrodes were prepared not only by chemical deposition of platinum, but also in the form of a wire or grid mechanically pressed onto a Nafion membrane. The main parameter that controls the behaviour of the Pt/Nafion electrode is the ratio of the true to the geometric area of the electrode surface (the roughness factor). The lower is this ratio, the more sensitive is the electrode potential to the changes in the hydrogen concentration in the gas phase. Hence, depending on the roughness factor value, the Pt electrode can be employed either as an indicator or a counter electrode of the hydrogen sensor:

Pt (H_2, air) | NAFION | Pt'(H_2, air)

where Pt and Pt' are platinum electrodes with different roughness factor values. (For these measurements, we have developed an electrolytic junction between an electrochemical sensor containing no liquid electrolyte and a classical second-kind reference electrode [17]. A strip of Nafion membrane - salt bridge - is pressed at one end to the solid polymer electrolyte in the sensor and the other end is immersed in an electrolyte solution in an auxiliary vessel containing the reference electrode. This experimental arrangement makes it possible to measure the potential of the individual sensor electrodes against a well-defined reference electrode.)

Based on the above mentioned cell, a simple solid-state hydrogen sensor has been constructed [18]. A platinum wire was used as an indicator electrode and the chemically deposited platinum as a counter electrode; the sensor structure is depicted in ***Fig. 4***. The electrodes attain different mixed potential values in air containing hydrogen because of their different roughness factors and thus the sensor does not require a reference gas. It can be used both for potentiometric (open-circuit potential is measured) and amperometric (electrodes are shorted through an A-meter) detection of hydrogen.

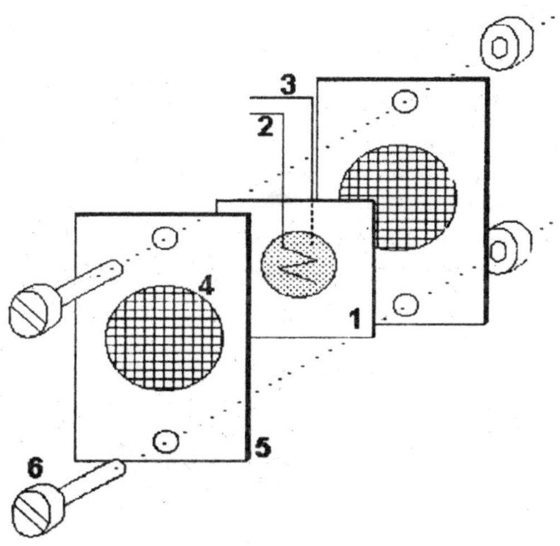

Fig. 4: **Hydrogen solid-state sensor with two different platinum electrodes**

1 - Nafion membrane, 2 - contact to the indicator electrode, 3 - contact to the counter electrode, 4 - supporting polyethylene grid, 5 - plexiglass body, 6 - assembling bolts

Tho detector sensitivity depends on the relative humidity (RH) of the test air. Within a hydrogen concentration range from 40 to 4000 ppm (v/v) and with RH varying between 33 and 95 %, the sensitivity of potentiometric detection decreases with increasing RH from 149 mV to 113 mV per concentration decade, while that for amperometric detection increases from 0.67 to 1.25 nA/ppm.

The dependence of the sensor sensitivity on RH clearly demonstrates that Nafion exchanges water with the environment and its physico-chemical properties are strongly dependent on the amount of water contained within the membrane. This phenomenon is of principal importance for the behaviour of all sensors containing Nafion as a solid electrolyte. It was studied many times, usually in an experimental arrangement securing that water content in Nafion does not change for the time of measurement, i.e. Nafion properties were obtained at certain equilibrium water contents [19]. We studied the changes in the Nafion properties (electric resistance) during the process in which one equilibrium content of water changed into another one due to exchange of water between Nafion and ambient air of a certain RH [20].

We have found out that uptake of water by Nafion membranes is much faster compared with the loss of water. The water uptake is governed by hydrophilic ions present in the membrane. These ions readily include water in their solvation mantles and thus facilitate entrance of water into Nafion. On the other hand, when the water content decreases, these ions form immobilized ion pairs that hinder transport of water from Nafion.

The rate of water transport is undoubtedly also affected by the cluster structure of Nafion. Changes in the humidity of ambient air primarily affect the state of the Nafion surface. On a decrease in the water content in Nafion, the clusters in the surface layer decrease in size, are reorganized and the connecting channels are narrowed. The surface layer is thus more difficult to penetrate and transport of water from Nafion bulk to its surface is hindered. On the other hand, on entrance of water into Nafion, the clusters are increased by the water molecules accepted and the connecting channels are widened, enhancing the water transport.

In electrochemical gas sensors, the electrochemical reactions take place at a three-phase boundary, i.e. at a place where gas, electrode and electrolyte come into contact with one another. If a grid electrode is formed, e.g. by vacuum evaporation of a metal onto a Nafion surface, the three-phase boundary is exactly defined. Its length equals the perimeter of the grid electrode. It was shown [21] that the gas sensor sensitivity increases with increasing perimeter, even if the total geometric area of the electrode decreases; the higher is the dimensionless ratio P^2/A, where P is length of the perimeter of the indicator electrode and A its geometric area, the higher is the signal/noise ratio.

Indicator electrodes vacuum-plated on the surface of a Nafion membrane are not mechanically stable. Therefore, we used an electroformed gold minigrid pressed onto a Nafion membrane as the indicator electrode in the solid-state amperometric sensor for detection of nitrogen dioxide in air [22]. The gold minigrid was sufficiently strong to withstand fluctuations in the membrane dimensions caused by humidity variations and its fine square grid structure ensured a high P^2/A ratio. Nitrogen dioxide was determined by reduction at -0.3 V *vs.* Pt/air electrode. The sensor sensitivity increased linearly with increasing RH of the test air; it was 59 nA/ppm NO_2 at RH = 42 %. The time constant and the time required to attain 95 % of the steady-state response were 2.2 and 10 s, respectively.

A minigrid electrode can be considered as an array of semimicro- or microband electrodes. The sensor response is then controlled to a certain extent by transport processes characteristic for microelectrode arrays; that is, the actual sensor sensitivity is much higher then expected for a sensor with a macroscopic electrode of the same geometric area as the total area of the array. The electrochemical sensor sensitivity thus depends not only on the catalytic activity of the indicating electrode material but, to a great extent, also on its geometric design.

6.3.2 Determination of substances in solution

Pneumatoamperometry (see above) is a very powerful method for determination of substances in solution that can be converted into equivalent amounts of volatile and electrochemically active products through chemical reactions with suitable reagents; this conversion improves the selectivity of the determination. When a MePME is employed for the volatile product detection, the membrane pores can be flooded by the electrolyte solution, resulting in a severe decrease in the sensor sensitivity. Therefore, a sensor employing a Nafion membrane coated with platinum on one side was used for pneumatoamperometric determination of nitrate and nitrite in waters and vegetables [23]. Nitrite and nitrate were chemically converted into nitrogen oxide which was detected in the gas phase on the platinum-coated side of the membrane. The other side of the Nafion membrane was in contact with an electrolyte solution containing an auxiliary and reference electrode. The limits of detection for NO_2^- and NO_3^- were 27 μg l^{-1} and 62 μg l^{-1}, respectively. The sample throughput was about 10 samples per hour.

6.4 Gas Sources for Calibration of Gas Sensors

It is necessary for laboratory testing and calibration of gas sensors to have a source of the test gas permitting simple variation of the test component concentration. Concentration changes of several orders of magnitude are often required, sometimes at a constant flow rate of the carrier gas. We developed and tested sources of some gases based both on permeation and on electrochemical generation.

6.4.1 Permeation Gas Sources

Permeation of the test gas through a polymeric material into a stream of a carrier gas is a method frequently used for the preparation of calibration mixtures. Classical permeation sources containing liquefied gas have some drawbacks - the relatively long time required for such source to produce a constant amount of gas and practical impossibility of changing gas production.

We developed a source (generator) of sulphur dioxide [24] based on permeation of SO_2 from a buffered (pH = 5.0) and thermostatted (30 °C) aqueous solution of sodium hydrogensulphite through the wall of a thin-walled silicone rubber tube into the stream of a carrier gas. In the generator, three (or more) permeation tubes of different length are simultaneously immersed in the generation solution. The tubes can be connected stepwise to the carrier gas source, so that three (or more) different concentration of SO_2 in the carrier gas at a constant flow rate can be obtained from a single generation solution.

The tested generator produced sulphur dioxide from 1.25 to 25.4 ng s^{-1}, depending on the hydrogensulphite concentration. The long-term stability was tested for the production of 10.5 ng s^{-1} over 50 hours; the relative standard deviation was 1.6 %. An equation was derived to estimate the SO_2 production for various solution compositions and surface areas of the permeation tubes. Low concentrations of SO_2 can be produced from concentrated hydrogensulphite solutions, so that the lifetime of the generator is sufficiently long (hundreds of hours) for laboratory applications. After the generator was filled and the solution thermostated, 30 minutes sufficed for the generator to produce a constant amount of SO_2. The general principle used can also be applied to the generation of other gases and vapours, e.g. hydrogen cyanide from a potassium cyanide solution [25].

A simple electronic thermostat for permeation sources, employing silicon power transistors as heaters, was tested [26] and an electromagnetically operated three-way valve permitting introduction of either a calibration gas or the test gas to the sensor, was also developed [27]. General aspects of construction, calibration and basic performance characteristics of permeation sources containing liquefied gas and those based on chemical generation of a gas in solution were discussed in the review [28].

6.4.2 Electrochemical Generation of Gases

A very suitable method for preparation of small amounts of gases is electrolytic (or coulometric) generation. Many gases can be generated electrolytically [29], usually at a constant current. This approach has a number of advantages:

- the gas is only generated when required;
- the rate of its production is readily controlled through the magnitude of the generating current;
- the gas production can be determined from Faraday's law for a given generating current (provided that the current efficiency is known);
- the gases which are not commonly available (AsH_3, SbH_3), dangerous gases (HCN) and some gases which cannot be liquefied under the permeation source conditions (CO, H_2) can be generated electrolytically.

Electrolytic generators of hydrogen cyanide, hydrogen and carbon monoxide were developed, tested and used for sensor calibration in our laboratory.

Hydrogen cyanide was generated [30] by constant-current oxidation of an aqueous solution of potassium thiocyanate at a platinum wire anode. The HCN produced was transferred into a carrier gas which passed through the platinum anode compartment. In a 0.1M KSCN solution the rate of production of HCN was a linear function of the generating current, from 10 to 200 μA. The relative standard deviation for an HCN production rate of 6 ng s^{-1} was 1.8 % and that for 0.9 ng s^{-1} was 5.9 %. The time required to establish a steady-state production after a change in the generating current was 10 minutes.

The hydrogen generator [31] was designed in such a way that the carrier gas only passed over the generating solution. Two construction variants were developed (***Fig. 5***), one having a very low dead volume. A Nafion membrane was used in the generator to separate the cathodic and anodic compartments. The membrane prevents the electrolyte solution (1M H_2SO_4) from being transported from one compartment to the other even if the pressure differs in the two compartments.

It is evident that the hydrogen generator described can also be used for oxygen generation; it was employed in a carbon monoxide generator [32]. The generation of CO is based on oxidation of carbon at an elevated temperature, 920-950 °C, by electrolytically generated oxygen. The oxidation takes place under conditions analogous to those for oxygen determination in elemental analysis of organic substances (the Unterzaucher method). The rate of the CO production was a linear function of the oxygen generating current over a range of 5 - 60 mA. It corresponded to the theoretical value calculated from Faraday's law assuming stoichiometry of $O_2 \rightarrow 2\ CO$. Obviously, only inert gases (nitrogen, argon, helium) not containing traces of oxygen, can be used as carrier gases with this generator.

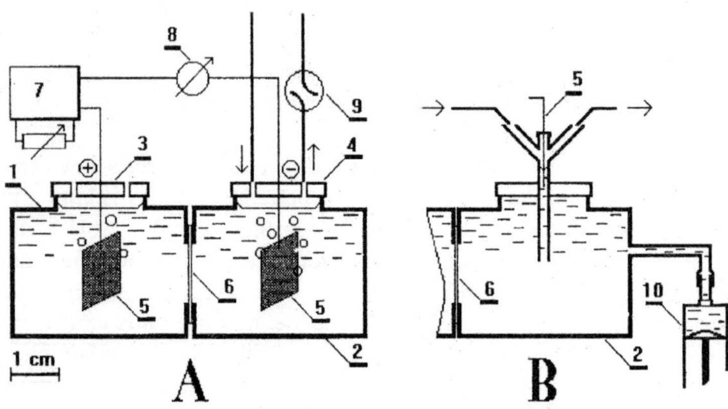

Fig. 5: **Electrolytic hydrogen generator**

1, 2 - vessels with an electrolyte solution, 3, 4 - lids, 5 - platinum electrodes, 6 - Nafion membrane, 7 - galvanostat, 8 - current meter, 9 - separator of the electrolyte solution droplets, 10 - syringe for adjusting of the electrolyte meniscus height in the generating space

6.5 Miscellaneous

A special enrichment unit already mentioned [13] has been employed in a wall-jet conductometric detector for continuous determination of sulphur dioxide at a ppb level in atmosphere [33]. Sulphur dioxide is preconcentrated from air into a polydisperse aerosol of water. The aerosol condenses on impact on the detector, on the surface of which a film of the condensate is maintained. The conductance of the liquid film is measured with a pair of platinum electrodes. The calibration dependence is not linear, but corresponds to the theoretical model. The concentration range of 19-800 ppb (v/v) was studied. The detection limit was 0.8 ppb, the relative standard deviation for 145 ppb was 2.4 %. A steady-state response to a change in the SO_2 concentration was attained within ca. 70 s. Although conductometric detection is inherently nonselective, the determination of sulphur dioxide in the atmosphere is not significantly affected by the presence of common gaseous pollutants because they are mostly present as undissociated molecules, except for ammonia [34]; ammonia is the only serious interferent.

Sulphur dioxide can be absorbed in the enrichment unit into an aerosol of buffered formaldehyde solution [35]. The solution from the enrichment unit is then analyzed spectrophotometrically employing pararosaniline method. A limit of determination was 2 ppb. A sample for spectrophotometric detection is available each 10 minutes under the conditions applied.

A survey of various electrochemical methods for the determination of sulphur-containing gases and sulphur-containing compounds transferable to the gas phase by a chemical reaction or by the reductive or oxidative pyrolysis was published in ref. [36]. In ref. [37], a survey of the basic principles of potentiometric determination of gaseous compounds is given with emphasis on methods based on the membrane gas sensors.

6.6 References

1. Janata J.: *Principles of Chemical Sensors*. Plenum Press, New York 1989.
2. Riley M. in: *Ion-selective Electrode Methodology, Vol.2* (Covington A. K., Ed.). CRC Press, Boca Raton 1979.
3. Opekar F.: in: *Instrumentation in Analytical Chemistry, Vol. II* (Zýka J., Ed.). Ellis Horwood, Chichester 1994.
4. Azad A. M., Akbar S. A., Mhaisalkar S. G., Birkefeld L. D., Goto K. S.: J. Electrochem. Soc. *139*, 3690 (1992).
5. Janata J., Josowicz M., DeVaney D. M.: Anal. Chem. *66*, 207R (1994).
6. Opekar F.: Electroanalysis *1*, 287 (1989).
7. Opekar F.: Chem. Listy *79*, 703 (1985).
8. Langmaier J., Opekar F., Pacáková V.: Talanta *34*, 453 (1987).
9. Langmaier J., Polák J., Opekar F.: Analyst *113*, 501 (1988).
10. Opekar F.: Anal. Chim. Acta *183*, 293 (1986).
11. Langmaier J., Opekar F.: Collect. Czech. Chem. Commun. *51*, 2077 (1986).
12. Opekar F., Večeřa Z., Janák J.: Intern. J. Environ. Anal. Chem. *27*, 123 (1986).
13. Večeřa Z., Janák J.: Anal. Chem. *59*, 1494 (1987).
14. Trojánek A., Opekar F.: J. Electroanal. Chem. *214*, 125 (1986).
15. Opekar F.: J. Electroanal. Chem. *260*, 451 (1989).
16. Opekar F., Langmaier J., Samec Z.: J. Electroanal. Chem. *379*, 301 (1994).
17. Opekar F.: Sensors and Actuators *B21*, 131 (1944).
18. Samec Z., Opekar F., Crijns G. J. E. F.: Electroanalysis, *in press*.
19. Zawodzinski T. A., Neeman M., Sillerud L. O., Gottesfeld S.: J. Phys. Chem. *95*, 6040 (1991).
20. Opekar F., Svozil D.: J. Electroanal. Chem. *385*, 269 (1995).

21. Buttner W. J., Maclay G. J., Stetter J. R.: Sensors and Actuators *B1*, 303 (1990).
22. Opekar F.: Electroanalysis *4*, 133 (1992).
23. Langmaier J., Opekar F.: Chem. Listy *84*, 80 (1990).
24. Langmaier J., Opekar F.: Anal. Chim. Acta *166*, 305 (1984).
25. Opekar F.: Chem. Listy *77*, 884 (1984).
26. Opekar F.: Chem. Listy *85*, 203 (1991).
27. Opekar F., Křesťan L.: Chem. Listy *82*, 189 (1988).
28. Opekar F.: Chem. Listy *84*, 128 (1990).
29. Hersch P., Sambucetti C. J., Deuringer R.: Chim. Anal. (Paris) *46*, 31 (1964).
30. Tocksteinová Z., Opekar F.: Talanta *33*, 688 (1986).
31. Opekar F.: Chem. Listy *88*, 258 (1994).
32. Opekar F., Langmaier J.: Talanta *39*, 368 (1992).
33. Opekar F., Trojánek A.: Anal. Chim. Acta *203*, 1 (1987).
34. Symanski J. S.,Bruckenstein S.: Anal. Chem. *58*, 1771 (1986).
35. Opekar F., Trojánek A.: Chem. Listy *81*, 649 (1987).
36. Langmaier J., Opekar F.: Chem. Listy *82*, 897 (1988).
37. Opekar F.: Chem. Listy *87*, 396 (1993).

7

ELECTROCHEMISTRY OF ENVIRONMENTALLY IMPORTANT ORGANIC SUBSTANCES

Jiří Barek and Jiří Zima

Abstract

The polarographic and voltammetric behaviour of environmentally important organic substances (selected chemical carcinogens, pesticides and dyes), studied at UNESCO Laboratory of Environmental Electrochemistry in the past decade, is reviewed and possible role of electrochemistry in elucidation of their genotoxic and ecotoxic properties, the mechanism of their action, metabolism, the fate in the environment etc., is briefly discussed. The use of modern electroanalytical techniques, namely differential pulse polarography and voltammetry, adsorptive stripping voltammetry and high-performance liquid chromatography with electrochemical detection, for the determination of trace amounts of these substances is described.

Key Words

differential pulse polarography (voltammetry), adsorptive stripping voltammetry, high-performance liquid chromatography with electro-chemical detection, chemical carcinogens, pesticides, dyes

7.1 Introduction

Many potentially and actually hazardous organic compounds are emitted from various anthropogenic sources into the environment. Obviously, they should be carefully monitored in order to introduce appropriate environmental protection regulations and induce ecochemical and ecotoxicological research to clarify the fate, behaviour and effects of these substances in various environmental compartments (atmosphere, hydro-sphere, terrestrial ecosystems, etc.). Therefore, there is an ever increasing demand for analytical methods suitable for the determination of trace amounts of various biologically active organic compounds in evironmental samples.

Modern polarographic and voltammetric methods, e.g. differential pulse polarography (DPP) [1-4], linear scan voltammetry (LSV) [1-3], adsorptive stripping voltammetry (AdSV) [1-3, 5-9], meet the stringent conditions on selectivity and sensitivity essential in the determination of these substances in the environment and in the study of their biotransformation. In many cases, it is necessary to use mixtures of aqueous buffers with suitable organic solvents to enhance the solubility of the test substances. For the determination of extremely low concentrations of many biologically active organic compounds, highly diluted base electrolytes yield smoother base-lines due to lower concentrations of trace impurities. This makes it possible to attain much lower limits of determination (LOD), especially with AdSV where it is advisable to decrease the concentration of organic solvent as much as possible, because its presence usually decreases the tendency of organic substances to adsorb on the HMDE surface. In many cases, extremely high sensitivity of modern polarographic and voltammetric techniques can be combined with highly efficient liquid chromatographic separation [10]. The theoretical background and many practical applications of these methods are described for environmental [11], clinical [5] and biological [12,13] sciences. Extensive information on the mechanisms of the electrode reactions of these substances can be found in ref. [14].

Moreover, the study of polarographic or voltammetric behaviour of biologically active organic compounds can provide a great amount of information about their electron-transfer reactions. Both electrochemical and biological reactions are essentially heterogeneous processes occurring at the electrode-solution or enzyme-solution interface. They can occur at similar pH values and temperatures and in both cases it is likely that the substrate molecule has to be oriented in a rather specific fashion before the electron transfer can occur. In our opinion, there is sufficient similarity between electrochemical and biological reactions to warrant extensive study of the electrochemistry of biologically active organic molecules, which should considerably contribute to elucidation of the fundamentals of the biological reaction mechanisms.

Therefore, the polarographic reduction of a number of different biologically active organic compounds was studied in detail in our laboratory over the past decade to establish the reduction mechanism and to find out optimal conditions for their polarographic and voltammetric determination. The electrode mechanism proposed was usually based on detailed investigation of the influence of the pH and the base electrolyte composition on DC or tast polarographic behaviour of the test compounds, cyclic voltammetric experiments, constant-potential coulometric determination of the number of electrons exchanged under different conditions and identification of the electrode reaction products using UV spectrophotometry, HPLC or TLC, DC voltammetry at a glassy carbon rotating disk electrode (RDE), etc.

This chapter is aimed at demonstrating that modern polarographic and voltammetric techniques can be successfully used for the determination of trace amounts of various environmentally important organic substances. Therefore, electrochemical behaviour of selected chemical carcinogens, herbicides, dyes and optical brightening agents is described, together with applications to environmental analysis.

7.2 Electrochemistry of Chemical Carcinogens

Cancer exacts substantial costs in treatment and preventive measures on a world-wide scale and causes immeasurable human suffering. The approximately 20 % cancer mortality, together with the fact that environmental causes contribute to a majority of cancers, emphasize the benefits of environmental detection of chemical carcinogens and makes the monitoring of carcinogens in general and working environment the highest priority [15]. Analytical measurement procedures should have a critical role in molecular epidemiology and exposure regulation, as well as in environmental monitoring [16].

In many cases, modern electroanalytical methods can be a viable alternative to more frequent spectrometric or separation methods. Moreover, in some cases there is a relationship between polarographic and/or voltammetric behaviour and genotoxic properties of organic compounds and the knowledge of the mechanism of their electrode reactions can provide a useful clue in elucidation of the mechanism of their interaction with living cells and their fate in the environment.

For the above reasons, we have systematically investigated polarographic and voltammetric behaviour of several groups of chemical carcinogens. The results of these investigations are briefly summarized in the following paragraphs.

7.2.1 Derivatives of N,N-dimethyl-4-aminoazobenzene

Derivatives of N,N-dimethyl-4-aminoazobenzene are suspected chemical carcinogens. Even very small amounts of these substances can have a detrimental effect on biological processes and thus a great deal of attention has recently been paid to analytical methods useful for determination of their traces in both biological samples and in working or natural environment.

In the past decade, we have thoroughly investigated polarographic behaviour of unsubstituted N,N-dimethyl-4-aminoazobenzene [17], its

3'-halogen [18], 3'-methyl [19], 3'-nitro [20], 4'-amino [21] and 4'-halogen [22] derivatives in mixed aqueous-methanolic medium. With increasing pH, the $E_{1/2}$ values of these substances were shifted towards negative potentials (see ***Fig. 1***).

This can be explained by a preceding protonation of the azo group causing lowering of the electron density in the region of the double bond between the nitrogen atoms which facilitates the electron transition from the electrode to the studied molecule. The observed pH dependence of the limiting current (see ***Fig. 1***) indicates that, in an acid medium, four electrons are transferred, whereas in an alkaline medium only two electrons are exchanged. CPC demonstrated transfer of four electrons over the whole pH range. Both these facts can be explained in terms of irreversible disproportionation of a hydrazo compound formed in the first step of the electrochemical reduction according to Eq. (A) and (B).

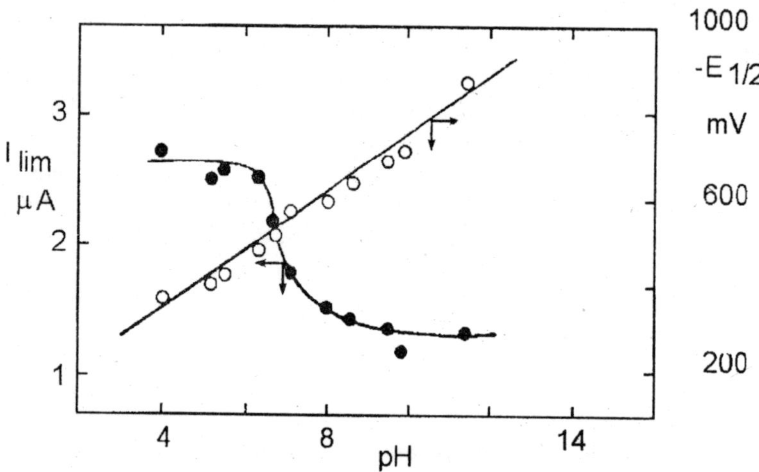

Fig. 1: **The dependence of I_{lim} and $E_{1/2}$ of N,N-dimethyl-4-amino-4'-fluoroazobenzene on the pH in a BR-MeOH (1:9) medium**

$$(CH_3)_2N-C_6H_4-N{=}N-C_6H_5 + 2H^+ + 2e^- \longrightarrow (CH_3)_2N-C_6H_4-NH{-}NH-C_6H_5 \qquad (A)$$

$$2\,(CH_3)_2N-C_6H_4-NH{-}NH-C_6H_5 \longrightarrow (CH_3)_2N-C_6H_4-N{=}N-C_6H_5 +$$

$$+ (CH_3)_2N-C_6H_4-NH_2 + H_2N-C_6H_5 \qquad (B$$

If the disproportionation occurs during the mercury drop lifetime, then the regenerated azo compound can undergo further reduction leading to transfer of four electrons. The decrease in the limiting current value to about one half, occurring in an alkaline medium, can be explained by a lower rate of disproportionation. At a sufficiently high pH, the half time of this reaction is probably one order of magnitude larger than the drop time but it is still shorter than the duration of the coulometric measurement which could explain the difference in the number of electrons exchanged, as found by tast polarography and CPC.

Polarographic behaviour of N,N-dimethyl-4-amino-3'-nitroazobenzene is more complicated because of the presence of nitro group which can be reduced with the exchange of 2, 4 or 6 electrons to -NO, -NHOH or $-NH_2$ group, respectively, in dependence on the pH [20]. Similarly, the mechanism of the reduction of N,N-dimethyl-4-amino-4'-amino-azobenzene is affected by the mutual interaction between the azo and amino group [21].

Optimum conditions for analytical utilization of these processes are summarized in ***Table 1***. The selectivity of these methods enables direct determination of N,N-dimethyl-4-aminoazobenzene in blood plasma [17] or analysis of a mixture of azobenzene, N,N-dimethyl-4-aminoazobenzene and N,N-dimethyl-4-amino-4'-aminoazobenzene by DPP [21]. Moreover, the selectivity can further be improved by a preliminary separation, e.g. by extraction into chloroform [17] or diethyl ether [19] or by TLC [19,21]. Anodic DPV at a glassy carbon RDE was found to be useful not only for the determination of the initial derivatives of N,N-dimethyl-4-aminoazobenzene but also for the products of their reductive splitting [23].

Tab. 1: **Polarographic and voltammetric determination of N,N-dimethyl-4-aminoazobenzene derivatives**

Substituent	*Technique*	*Medium*	*LOD*	*Ref.*
H	DPP	NH_3-NH_4Cl-MeOH (1:1) pH 9.7	1.10^{-8} M	17
	AdSV	NH_3-NH_4Cl-MeOH (1:1) pH 9.7	3.10^{-9} M	
3'-CH_3	DPP	BR-MeOH (1:9) pH 9.6	2.10^{-8} M	19
	AdSV	BR-MeOH (1:9) pH 4.6	3.10^{-9} M	
4'-NH_2	DPP	BR-MeOH (1:9) pH 9.7	2.10^{-8} M	21
	AdSV	BR-MeOH (1:9) pH 4.6	2.10^{-8} M	
4'-X	DPP	BR-MeOH (1:9) pH 6.7	5.10^{-8} M	22
(X=F,Cl,Br,I)	AdSV	BR-MeOH (1:1) pH 5.5	1.10^{-8} M	
3'-X	DPP	BR-MeOH (1:9) pH 6.1	2.10^{-7} M	18
(X=F,Cl,Br,I)	AdSV	BR-MeOH (1:9) pH 2.8	2.10^{-9} M	
4'-NO_2	DPP	BR-MeOH (9:1) pH 2.1	1.10^{-7} M	20
	AdSV	BR-MeOH (4:6) pH 8.4	2.10^{-9} M	

7.2.2 N-nitrosocompounds

N-nitrosocompounds are among the most commonly occurring chemical carcinogens and are present in trace amounts in a number of products of the chemical and even food industries, in various agricultural products and also in the natural and working environment.

Derivatives of N-nitroso-N-methylaniline constitute an interesting group of substances with an unusual organ specificity; they almost exclusively cause cancer of the pharynx. We have investigated polarographic behaviour of the 4-substituted derivatives of N-nitroso-N-methylaniline (R = H, CH_3, OCH_3, Cl, CN, OH and NO_2) in a mixed BR-MeOH (1:1) medium at pH 2-13 [24]. In an acidic medium, these substances undergo four-electron reduction according to Eq. (C), while a two-electron reduction occurs in an alkaline medium according to Eq. (D).

$$R\text{-}C_6H_4\text{-}N(NOH^+)(CH_3) \xrightarrow{4\,e^-,\,4\,H^+} R\text{-}C_6H_4\text{-}N(NH_3^+)(CH_3) + H_2O \qquad \text{(C)}$$

$$R\text{-}C_6H_4\text{-}N(NO)(CH_3) \xrightarrow{2\,e^-,\,2\,H^+} R\text{-}C_6H_4\text{-}N(NHOH)(CH_3) \qquad \text{(D)}$$

The E_p values for the individual substances are basically determined by the electron donor properties of the substituent in the 4-position. ***Fig. 2*** depicts an acceptable correlation between E_p and the Hammett constants σ_p of the corresponding substituents, confirming a single mechanism for the reduction of the test series of substances.

It was demonstrated that FSDPV at a HMDE gives linear concentration dependence in the concentration range, 1×10^{-5}-1×10^{-7} mol l^{-1} , and that this technique can be used for the analysis of a mixture of the test substances either directly or after a TLC separation. False positive results frequently occurring in environmental analysis can be eliminated by UV irradiation of the test sample leading to denitrosation of N-nitrosocompounds and yielding polarographically inactive products.

Conditions have been optimized for the reversed-phase HPLC separation of these substances with a C18 chemically bonded stationary phase [25]. Four detection techniques have been investigated: UV photometry, DCV at a HMDE, anodic voltammetry at a glassy carbon fibre array detector and an indirect anodic voltammetric detection after photolytic denitrosation of the analytes yielding corresponding aromatic amines. The latter technique has limits of detection around 1×10^{-7} mol l^{-1}, i.e. one order of magnitude lower than those obtained UV photometrically.

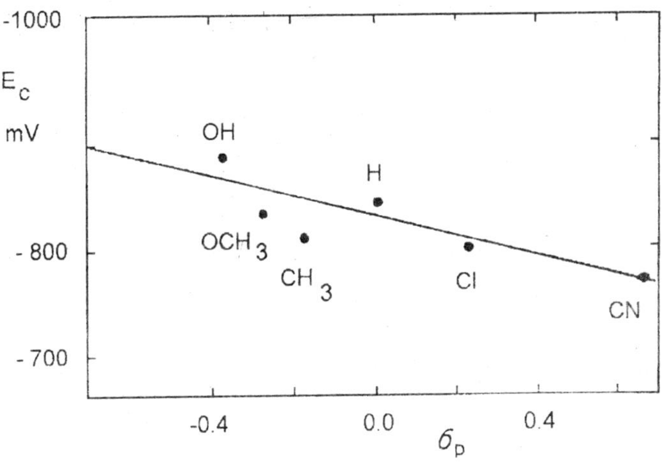

Fig. 2: **Dependence of E_p of 4-R-N-nitroso-N-methylaniline on the Hammett constants σ_p of substituent R in a BR-MeOH (1 : 1) medium at pH 2.7**

N,N'-dinitrosopiperazine is a suspect chemical carcinogen frequently occurring in trace amounts in various components of the working and living environment. We have found [26] that at pH 10 this substance is polarographically reduced according to Eq. (E), while at pH 2.4 an eight-electron reduction occurs, according to Eq. (F).

$$ON{-}N(CH_2CH_2)_2N{-}NO \xrightarrow[3H_2O]{4e^-} HN(CH_2CH_2)_2NH + N_2O + 4\,OH^- \qquad (E)$$

$$ON{-}N(CH_2CH_2)_2N{-}NO \xrightarrow{8e^-,\ 8H^+} H_2N{-}N(CH_2CH_2)_2N{-}NH_2 + 2\,H_2O \qquad (F)$$

CV on HMDE revealed that the process is irreversible irrespective of the pH. Optimum conditions (BR at pH 2.4) were established for the determination of this substance by DPP or FS DPV over a concentration range of 1×10^{-3} - 1×10^{-7} mol l^{-1}. The methods are applicable to the monitoring of the chemical efficiency of destruction of the test substance based on its oxidation with potassium permanganate.

Attempts at increasing the sensitivity of the voltammetric determination of all the test N-nitrosocompounds via their adsorptive accumulation on the HMDE surface failed.

7.2.3 Derivatives of 1-phenyl-3,3-dimethyltriazene

Derivatives of 1-phenyl-3,3-dimethyltriazene are among genotoxic substances which act via an alkylation mechanism. At the same time, they exhibit carcinostatic properties. With regard to easy polarographic reducibility of the triazene group, we have been concerned with applicability of modern polarographic and voltammetric techniques to the determination of trace amounts of these substances [27-33].

1-(2'-carbamoylphenyl)-3,3-dimethyltriazene yields two waves in a BR-MeOH (1:1) medium [27]. The $E_{1/2}$ of the first wave shifts to more negative potentials with increasing pH but this dependence is asymptotic rather than linear. At pH < 4 this wave decreases with time which is apparently connected with the decomposition of the test substance according to Eq. (G). The height of the first wave decreases at pH > 6 and completely disappears at pH > 10. Simultaneously, a new wave appears at pH > 6; its height and position are pH-independent. We have proved that the test substance is reduced according to Eq. (H), both in acidic and in alkaline media.

$$\text{C}_6\text{H}_4(\text{CONH}_2)\text{–N=N–N(CH}_3)_2 \xrightarrow{H^+} [\text{C}_6\text{H}_4(\text{CONH}_2)\text{–N}\equiv\text{N}]^+ + \text{HN(CH}_3)_2 \longrightarrow \text{(cyclic product, C=O, NH, N=N)} + H^+ \quad \text{(G)}$$

$$\text{C}_6\text{H}_4(\text{CONH}_2)\text{–N=N–N(CH}_3)_2 \xrightarrow[\text{IRREV}]{4\,e^-,\ 4\,H^+} \text{C}_6\text{H}_4(\text{CONH}_2)\text{–NH}_2 + \text{H}_2\text{N–N(CH}_3)_2 \quad \text{(H)}$$

The first reduction step yields the radical anion, which is immediately stabilized in an acidic medium by accepting a readily available proton, resulting in a dependence of $E_{1/2}$ on the pH. In contrast, in an alkaline medium, where there is a low concentration of free hydroxonium ions, this anion radical is stabilized by a proton that is removed from a water molecule, so that $E_{1/2}$ is pH-independent within this region. The observed atypical shapes of tast polarograms in the pH region 7-9 are apparently connected with the presence of MeOH and its effect on the preliminary protonation reaction. We assume that a substance adsorbed on the electrode surface undergoes preliminary protonation. The observed decrease in the limiting current at pH 7-9 in the region between -1.2 and -1.5 V is apparently connected with a decrease in the surface concentration of the adsorbed triazene at potentials far off from electrocapillary zero, where maximum adsorption of uncharged triazene molecules can be expected. This leads to a substantial decrease in the rate of the surface protonation reaction resulting in the decrease in the observed limiting current.

The mechanism of the polarographic reduction of 1-(3'-carbamoyl-phenyl)-3,3-dimethyltriazene [28], 1-(4'-carbamoylphenyl)-3,3-dimethyl-triazene [29] and 1-(4'-bromophenyl)-3,3-dimethyltriazene [30] is similar. However, the transient product of the acid-catalyzed decom-position of these substances cannot be stabilized by intramolecular cyclization and is further decomposed to form a polarographically inactive product.

Polarographically interesting is the behaviour of triazene derivatives carrying an additional, polarographically active substituent that can contribute to the sensitivity of their determination, as is the case with

1-[4'-(phenylazo)-phenyl]-3,3-dimethyltriazene [31] and 1-(2'-nitrophenyl)-3,3-dimethyltriazene [32]. The mechanism of the polarographic reduction of these substances is clarified in the cited papers.

Optimum conditions, under which the highest, best developed and most easily evaluated DPP or AdSV peaks can be obtained, are summarized in ***Table 2***.

Tab. 2: **Polarographic and voltammetric determination of some derivatives of 1-phenyl-3,3-dimethyltriazene**

Substituent	Technique	Medium	LOD	Ref.
2'-$CONH_2$	DPP/SMDE	BR-MeOH (9:1) pH 4.1	1.10^{-7} M	27
	AdSV	BR-MeOH (99:1) pH 5.1	3.10^{-9} M	
3'-$CONH_2$	DPP/DME	BR-MeOH (1:1) pH 5.9	2.10^{-7} M	28
	AdSV	BR-MeOH (99:1) pH 5.1	4.10^{-9} M	
4'-$CONH_2$	DPP/DME	BR-MeOH (1:1) pH 5.8	2.10^{-7} M	29
	AdSV	BR-MeOH (999:1) pH 4.0	4.10^{-9} M	
2'-NO_2	DPP/DME	BR-MeOH (1:1) pH 11.1	1.10^{-7} M	32
	AdSV	BR-MeOH (1:1) pH 7.2	1.10^{-8} M	
4'-C_6H_4-N=N-	DPP/DME	BR-MeOH (1:1) pH 8.0	1.10^{-7} M	31
	AdSV	BR-MeOH (999:1) pH 8.2	2.10^{-11}M	33
4'-Br	DPP/DME	BR-MeOH (9:1) pH 5.1	5.10^{-8} M	30

7.2.4 Some other chemical carcinogens

4-Nitrobiphenyl is suspect of chemical carcinogenity, as it can be enzymatically reduced to the proven chemical carcinogen 4-aminobiphenyl. Because of easy reducibility of the nitro group, modern polarographic and voltammetric methods can be used for its extremely sensitive determination [34, 37]. Because of the presence of an aromatic system, this analyte is strongly adsorbed on the HMDE surface, which can decrease LOD via adsorptive accumulation (see ***Table 3***).

Acridine, benzoacridine and dibenzoacridine derivatives are among chemical carcinogens frequently occurring in air due to diverse combustion processes. They have also been found in tobacco smoke, thermally treated food with high protein contents and in sea water and sediments in the vicinity of industrial centres. Heteroaromatic compounds with a π-electron deficit, which include the above mentioned acridine derivatives, are characterized by low energies of the lowest occupied molecular orbital and thus by polarographic reducibility. Moreover, their polarographic reduction can further be facilitated by protonation of the nitrogen atom, which brings about an additional decrease in the electron density at the sites of potential electron transfer. The addition of another aromatic ring to acridine results in an energy increase for the lowest occupied molecular orbital. This is naturally mirrored by an $E_{1/2}$ shift to more negative values. Due to a limited solubility of the test substances in water, aqueous-methanolic solutions must be used for their polarographic determination [35]. As expected, dibenz[a,h]-acridine accumulates on the HMDE surface, so that AdSV can be used for its determination (see ***Table 3***).

Benzyl chloride is an effective alkylation agent whose reactivity towards cellular nucleophiles leads to a high genotoxicity. C-Cl bond is rather difficult to reduce polarographically. Therefore, we had to use TBABr in DMF as the base electrolyte. In this medium, benzyl chloride yields one well developed, strongly irreversible, diffusion controlled wave with $E_{1/2}$ value -2.25 V *vs.* SCE [36]. Because of this highly negative potential, modern polarographic methods yield relatively high LOD values (see ***Table 3***).

Tab. 3: **Polarographic and voltammetric determination of some other chemical carcinogens**

Substance	Technique	Medium	LOD	Ref.
4-Nitrobiphenyl	DPP/DME	BR-MeOH (1:1) pH 8.6	6.10^{-8} M	37
	AdSV	BR-MeOH (1:1) pH 4.8	2.10^{-9} M	34
Acridine	AdSV	BR-MeOH (1:1) pH 6.0	2.10^{-8} M	35
Benz[c]acridine	AdSV	BR-MeOH (1:1) pH 4.1	5.10^{-9} M	35
Dibenz[a,h]acridine	AdSV	BR-MeOH (1:9) pH 4.8	2.10^{-8} M	35
Benzyl chloride	DPP/DME	0.05M TBABr in DMF	2.10^{-6} M	36

7.3 Destruction of chemical carcinogens and monitoring of its efficiency using modern electroanalytical techniques

Research on the physical, chemical, biological and environmental properties of chemical carcinogens inevitably results in the production of laboratory wastes containing these substances and may lead to contamination of the laboratory and equipment. Health safety regulations thus require that efficient methods be developed for destruction of these substances and for decontamination of the laboratory, equipment and waste. These methods must be efficient, simple, fast and cheap and must be combined with a suitable analytical method for controlling the chemical efficiency of the decontamination. Therefore, the International Agency for Research on Cancer (IARC) established a special programme to develop an authoritative series of monographs on methods for the destruction and disposal of carcinogenic waste from research laboratories. The project is based upon the experience of the IARC in bringing together internationally recognized experts to review data applicable to the destruction and disposal of carcinogenic waste, decontamination of research and analytical laboratories, to recommend destruction strategies, develop new methods and subject the designated methods to interlaboratory collaborative verification of their efficacy.

The UNESCO Laboratory of Environmental Electrochemistry took part in this project and a number of methods suitable for destruction of chemical carcinogens, decontamination of research and analytical laboratories and monitoring the efficacy of decontamination and/or destruction were developed in the past decade (see ***Table 4***).

Tab.4: **Methods of destruction of chemical carcinogens and analytical methods for monitoring their efficiency developed at the UNESCO Laboratory of Environmental Electrochemistry**

Carcinogen	Method of destruction	Analytical methods	Ref.
Acridine derivatives	$KMnO_4/H_2SO_4$	DPV/HMDE	35, 48
Antineoplastic pharmaceuticals	$KMnO_4/H_2SO_4$	DPP/DME	38, 42, 49
4-Nitrobiphenyl	Zn/HAc + $KMnO_4/H_2SO_4$	DPP/DME+DPV/RDE	39
Melphalan	$KMnO_4$/NaOH	DPV/RDE	40
Dimethylamino-azobenzene derivatives	$KMnO_4/H_2SO_4$	DPP/DME	41
Aromatic Amines	$KMnO_4/H_2SO_4$	DPV/RDE, HPLC-ED	43-47, 50
Mycotoxines	$KMnO_4$/NaOH or NaClO	HPLC	51

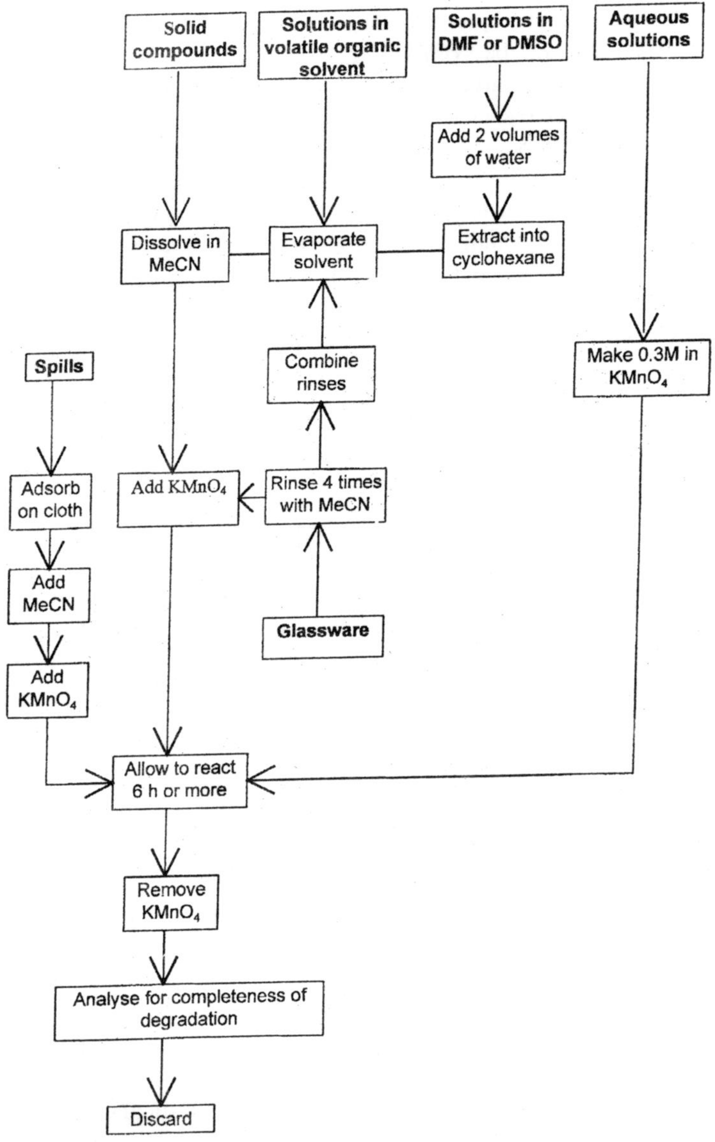

Fig. 3: **Schematic representation of the procedure for destruction of some polycyclic heterocyclic hydrocarbons**

The chemical efficiency of the destruction and/or decontamination of these substances can be verified using modern polarographic and voltammetric techniques given in ***Table 4***. The biological efficiency was checked by testing the residues produced by these methods for mutagenicity using Ames tests with *Salmonella typhimurium* strains TA 98, TA 100 and TA 102. All the methods have been tested collaboratively and afford a better than 99 % degradation. It can be expected that these methods can be used for a number of structurally similar substances. However, it is always necessary to verify the chemical and biological efficiency before using these techniques for untested substances.

Detailed procedures for destruction of solid substances, their solutions in various solvents, decontamination of spills of solid compounds, their solutions and of glassware are given in the cited IARC monographs. For the sake of illustration, a schematic representation of the procedure for destruction of dibenz[a,j]acridine, *7H*-dibenzo[c,g]carbazole and *13H*-dibenzo[a,i]carbazole is given in ***Fig. 3***.

It can be seen from ***Table 4*** that DPP/DME or DPV/HMDE is especially suited for monitoring the efficiency of destruction of polarographically reducible chemical carcinogens (e.g. those containing -NO_2, -NO, -N=N- functional groups or heteroaromatic systems). For polarographically inactive chemical carcinogens, such as aromatic amines or some pharmaceuticals, DCV or DPV at a glassy carbon RDE is applicable. HPLC-ED with a carbon fibre voltammetric detector is especially well suited for aromatic amines [52].

7.4 Electrochemistry of pesticides and growth stimulators

The analysis of pesticide residues has been the driving force behind modern ultratrace analysis. The application of voltammetric methods to the determination of agrochemicals (pesticides, growth stimulators etc.) in formulations, foodstuffs, crop residues, natural waters and body fluids has for long been recognized and practized in Eastern European countries [53, 54]. Examples of polarographically active pesticides are given in ***Table 5.***

Practical applications of modern polarographic and voltammetric methods for the determination of trace amounts of pesticides and growth stimulators require a suitable preliminary separation, clean-up and/or preconcentration. However, the extremely high sensitivity of these techniques make them very suitable for these purposes.

Some examples of new methods recently developed at the UNESCO Laboratory of Environmental Electrochemistry are summarized in ***Table 6***.

Pendimethalin yields a single polarographic wave at pH 1.3, with the value of $E_{1/2}$ = -0.25 V, which corresponds to the exchange of 12 electrons according to Eq. (I). It can be assumed that an ECE process with -NO and -NHOH intermediates is involved [56]. At pH 3-9, the stepwise reduction results in two separated waves. Under these conditions, pendimethalin is present in an unprotonated form, which is reduced after preceding protonation. This explains the observed dependence of $E_{1/2}$ on the pH.

***Tab. 5*: Examples of polarographically active pesticides**

Pesticide	Functional Group	Base electrolyte	$E_{1/2}$, V *vs.* SCE
Parathion	$-NO_2$	Ac, pH 5	-0.62
DDT	$-CCl_3$	TEABr	-0.73
Guthion	>C=O	Ac, pH 4	-0.83
Thiram	-S-S-	HCl	-0.26
Morestan		TBABr	-0.96
Diquat		KCl	-1.06

***Tab. 6*: Polarographic and voltammetric determination of some pesticides and growth stimulators**

Substance	Technique	Medium	LOD	Ref.
2-Nitrophenol	AdSV	BR, pH 5.0	3.10^{-8} M	55
4-Nitrophenol	AdSV	BR, pH 5.0	3.10^{-10} M	55
2,4-Dinitrophenol	AdSV	BR, pH 5.0	2.10^{-9} M	55
2-Methoxy-5-nitrophenol	AdSV	BR, pH 5.0	2.10^{-9} M	55
Pendimethalin	AdSV	BR-MeOH(9:1),pH 7	1.10^{-9} M	56
Chloridazon	AdSV	0.01M HCl	4.10^{-8} M	57

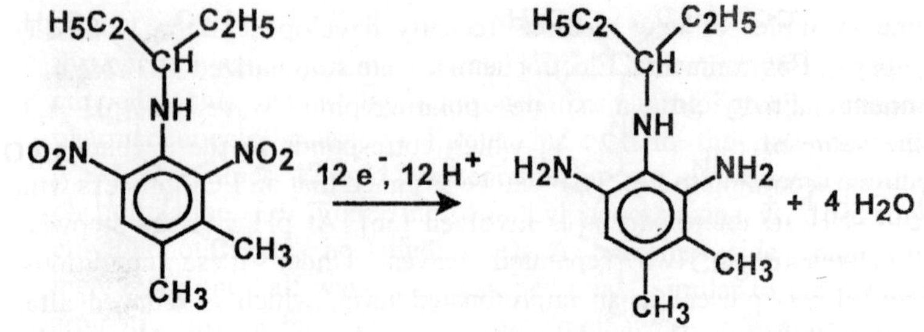

$$\text{pendimethalin} \xrightarrow{12\,e^-,\ 12\,H^+} \text{diamine product} + 4\,H_2O \quad \text{(I)}$$

Chloridazon is reduced with the exchange of 2 electrons, according to Eq. (J) [57]. The reduction of protonated and unprotonated species results in two separate polarographic waves and the product of the two-electron reduction undergoes a follow-up hydrolytic decomposition according to Eq. (K).

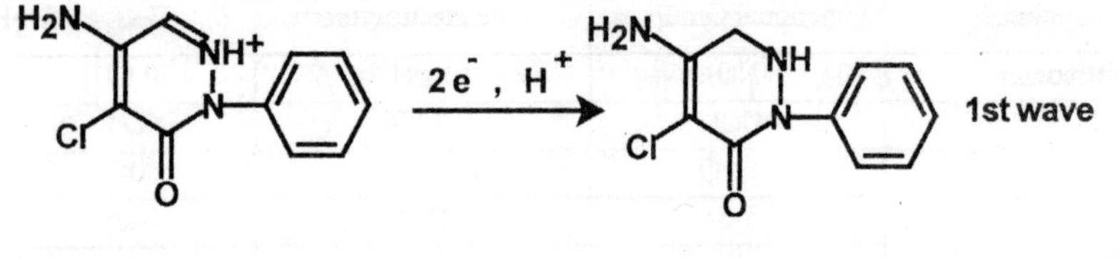

$$\xrightarrow{2e^-,\ H^+} \quad \text{1st wave}$$

$$\xrightarrow{2e^-,\ 2H^+} \quad \text{2nd wave} \quad \text{(J)}$$

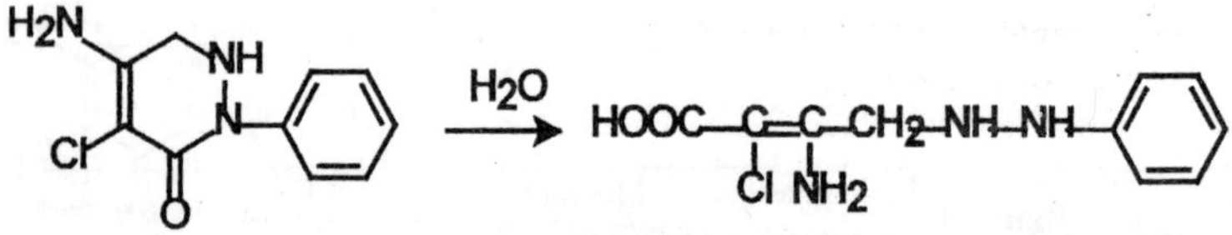

$$\xrightarrow{H_2O} HOOC{-}C(Cl){=}C(NH_2){-}CH_2{-}NH{-}NH{-}C_6H_5 \quad \text{(K)}$$

For determination of the above agrochemicals in environmental samples (e.g. drinking, river or surface waters), the DPP or AdSV technique must be combined with a suitable preliminary separation (extraction of chloridazon into dichloromethane or pendimethalin into dimethylether). HPLC-ED is well-suited for the analysis of mixtures of nitrated phenols in growth stimulators [57].

7.5 Electrochemistry of dyes and optical brightening agents

Azo dyes, which form the largest group of organic dyes, constitute more than 35 % of the global dye production and thus are frequently encountered in working and natural environment. A number of azo dyes exhibit genotoxic or ecotoxic properties, leading to the need for sensitive and selective methods of their determination. Because of easy reducibility of the azo group, modern polarographic and voltammetric methods are suitable for the determination of trace amounts of these substances in the production plant environment.

Optical brightening agents are another important group of dye industry products. Although modern types of these agents are not supposed to pose health or environmental hazards, some derivatives of coumarin are known to cause photosenzitization and exhibit mutagenic activity and some derivatives of 4,4'-diaminostilbene-2,2'-disulphonic acid are suspected chemical mutagens. Therefore, sensitive methods are needed for the determination of trace amounts of such substances, which come into the environment from households, laundries and textile and paper industries.

7.5.1 Azo dyes

In the past decade, we devoted attention to the polarographic and voltammetric determination of trace amounts of some genotoxic azo dyes (congo red [58], trypane blue [59], semitrypane blue [60]), several water-insoluble azopigments [61-65] and some optical brightening agents derived from coumarin [66,67] and 4,4'-diaminostilbene-2,2'-disulphonic acid [68,69]. The results of these papers are briefly summarized in the following paragraphs using congo red, trypane blue, 7-[4-methyl-5-phenyl-2-(1,2,3-triazolyl)]-3-phenylcoumarin and 4,4'-bis[(4-phenylamino-6-methoxy-1,3,5-triazin-2-yl)amino]stilbene-2,2'-disulphonic acid as examples. Further information on the polarographic behaviour of other investigated substances can be found in the cited papers.

Congo red (**I** in Eq. (L)) is one of the substances suspected of chemical carcinogenity, requiring two-step metabolic activation. *In vitro* and *in vivo* experiments have demonstrated that it is metabolically transformed into benzidine. However, the different organ specificity of congo red and benzidine indicates that the carcinogenic activity of this bisazodye is not dependent on its metabolites alone. Therefore, a detailed study of the polarographic reduction of congo red was carried out [58], in order to clarify the mechanism of the reduction and find out optimal conditions for its analytical utilisation.

We have found that at pH < 3, where the red form (**I**) is converted into the blue form (**II**), congo red is irreversibly reduced according to Eq. (L). The presence of benzidine (**III**) in the solution after coulometric reduction at a constant potential was demonstrated using UV spectrophotometry and TLC. At pH 6-8, a direct eight-electron reduction of the red form (**I**) occurs and at pH > 8, a four-electron irreversible reduction occurs according to Eq. (M).

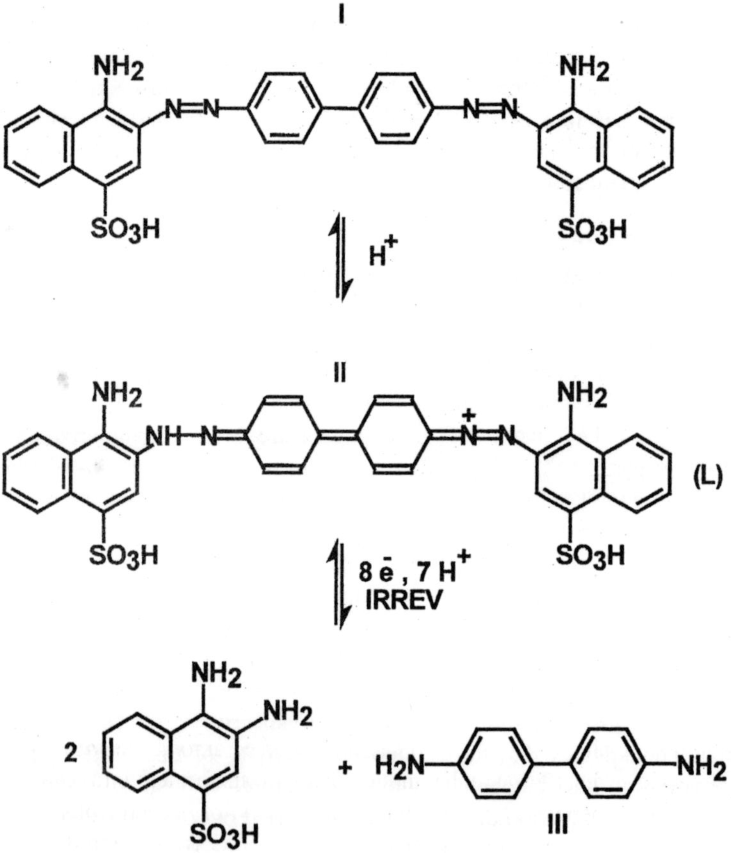

I

NH_2 NH_2 N=N N=N SO_3H SO_3H

H^+

II

NH_2 NH_2 NH–N= =$\overset{+}{N}$=N SO_3H SO_3H (L)

$8\bar{e}$, $7H^+$
IRREV

2 NH_2 NH_2 SO_3H + H_2N– –NH_2

III

$\rightleftharpoons$ $4\bar{e}, 4H^+$ (M)

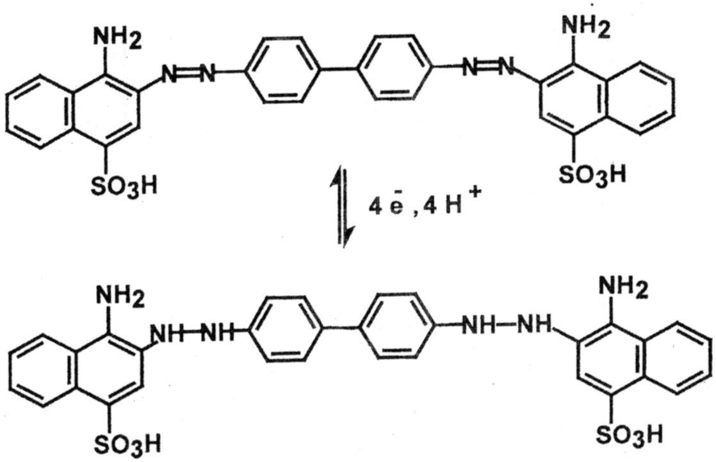

Optimum conditions for the determination of congo red are summarized in ***Table 7.***

Tab. 7: **Polarographic and voltammetric determination of selected azo dyes**

Substance	Method	Medium	LOD	Ref.
Congo red	DPP	BR, pH 7	2.10^{-8} M	58
	DPV	BR, pH 7	4.10^{-8} M	
	AdSV	BR, pH 7	2.10^{-9} M	
Trypane blue	DPP	0.1M H_3PO_4	8.10^{-8} M	59
	DPV	0.1M H_3PO_4	6.10^{-8} M	
	AdSV	0.1M H_3PO_4	5.10^{-9} M	
Semitrypane blue	DPP	0.1M NaOH	8.10^{-8} M	60
	DPV	BR, pH 8	5.10^{-9} M	
	AdSV	BR, pH 8	8.10^{-11} M	

Trypane blue (IV in Eq.(N)) can be classified among the substances with genotoxic effects. It can be metabolically transformed into carcinogenic 3,3'-dimethylbenzidine. The genotoxic activity of trypane blue makes it desirable that a number of independent methods be available to determine its low concentrations in biological and ecotoxicological studies.

In our detailed study of its polarographic behaviour [59] we have first investigated the influence of the pH. The observed shift of $E_{1/2}$ toward more negative values with increasing pH can be explained in terms of preceding protonation of the azo group, leading to the decrease in the electron density in the region of the double bond between the nitrogen atoms, facilitating the electron transfer. The irreversibility of the reducfion was confirmed by logarithmic analysis and CV at a HMDE. The coulometric reduction at a constant potential of -0.3 V in 0.1M H_3PO_4 indicated the exchange of eight electrons and 3,3'-dimethylbenzidine was identified as a reaction product using UV spectrophotometry and TLC. It can thus be assumed that trypane blue undergoes irreversible reduction according to Eq. (N) under the given conditions.

It can be seen from ***Table** 7* that the newly developed procedures can be used for the determination of extremely low concentrations of trypane blue and that the situation is analogous with structurally similar semitrypane blue, which was also investigated [60].

IV

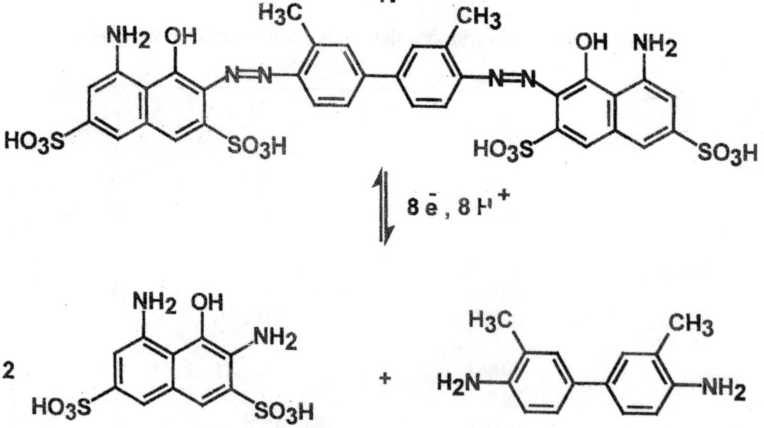

H3C CH3

NH2 OH OH NH2

N=N N=N

HO3S SO3H HO3S SO3H

8 e, 8 H+

(N)

NH2 OH

NH2

2 H3C CH3

\+ H2N NH2

HO3S SO3H

7.5.2 Optical brightening agents

Because of a low solubility of 7-[4-methyl-5-phenyl-2-(1,2,3-triazolyl)] -3-phenyl-coumarin (MPTPC, **V** in Eq.(O)) in water, BR-MeOH (1:1) mixtures were used as the base electrolyte. Over a range of pH 2-5, the $E_{1/2}$ values shift to more negative potentials with increasing pH, the slope of this linear dependence being 66 mV per pH unit. The $E_{1/2}$ values remain constant at pH > 7. The wave height is virtually constant within the pH 2-5 range. The waves are difficult to evaluate at pH < 3 because they are overlapped by the hydrogen reduction wave. At pH > 5, the MPTPC wave decreases, presumably due to alkaline hydrolysis according to Eq. (O).

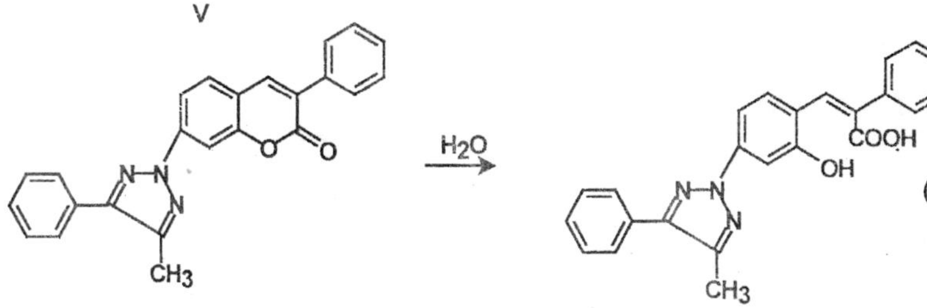

V

$\xrightarrow{H_2O}$ (O)

It has been confirmed that polarographic reduction is irreversible (logarithmic analysis and CV at a HMDE) and diffusion controlled (linear dependence of I_{lim} on $h^{1/2}$). The limiting current of MPTPC observed, within a range of pH 2-5, is nearly twice as high as that of unsubstituted coumarin, thus indicating a different mechanism of polarographic reduction. In view of the +M effect of the aromatic ring in position 3 on the coumarin ring, the MPTPC half wave potential should shift to more negative values compared to unsubstituted coumarin. The actual shift is opposite, which may be due to the fact that the phenyl ring in position 3 is not coplanar with the coumarin ring and thus conjugation cannot occur. The proposed mechanism of the polarographic reduction of MPTPC includes a two-electron reduction according to Eq. (P), analogous to the reduction of α,β-unsaturated carbonyl compounds. In view of the observed height of the MPTPC wave, the rate of dimerization of the transient radical **VI**, if occurring at all, is lower than that of consecutive reduction to the saturated carbonyl compound **VII.**

The BR-MeOH (1 : 1) mixture at pH 5 appears optimal from the analytical point of view. The dependence of the wave or peak height on the analyte concentration is linear over the region of 2-10 μmol l^{-1} for tast polarography (LOD = 2×10^{-6} mol l^{-1}) and 0.02-10 μmol l^{-1} for DPP (LOD = 5×10^{-8} mol l^{-1}). The electrocapillary curves of coumarin give evidence of strong adsorption of this substance at the DME surface. Therefore, we examined the possibility of using adsorptive accumulation of MPTPC at a HMDE and obtained LOD = 4×10^{-9} mol l^{-1} using AdSV at a HMDE with 60 sec accumulation from unstirred solution.

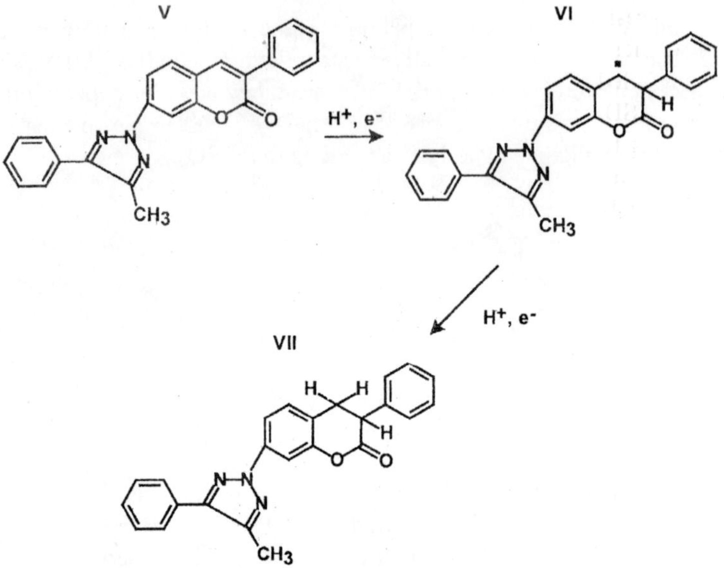

V

VI

H+, e-

(P)

H+, e-

VII

The newly developed methods can be used for the analysis of technical products. For waste water analysis, a suitable preliminary separation method must be used (e.g. extraction combined with TLC).

Two functional groups in the 4,4'-bis-[(4-phenylamino-6-methoxy-1,3,5-triazin-2-yl)-amino]stilbene-2,2'-disulphonic acid molecule (BASDA, **VIII** in Eq. (Q)) can play a role in its polarographic reduction, viz. triazine ring and the C=C bond of the stilbene system. Various stilbene derivatives yield waves with $E_{1/2}$ within a range of -2.1 to -2.8 V *vs.* SCE, in dependence on the medium. Therefore, DMF containing 5 % vol. of water was chosen as the solvent and TEABr was used as the supporting electrolyte. With this medium, it was possible to apply high negative potentials, required to reduce the double bond of the stilbene system in the BASDA molecule. (Nonaqueous DMF appeared unsuitable due to frequent distortions associated with irregular dropping of mercury. This drawback was eliminated by adding some quantity of water).

In the above medium, BASDA yields two waves. The first one ($E_{1/2}$ = -2.01 V), corresponds to the reduction of sodium ions, since BASDA is supplied as the disodium salt. This wave is analytically useless because practical samples always contain some inorganic salts which affect its height. Attention was, therefore, paid to the second wave ($E_{1/2}$ = -2.27 V) which apparently corresponds to the reduction of the stilbene double bond. This wave corresponds to an irreversible process and its limiting current is diffusion-controlled. We assume that it corresponds to a one-electron reduction of BASDA to the radical anion **IX** with subsequent acceptance of a proton according to Eq. (Q). The wave corresponding to the exchange of the second electron (a further reduction of free radical **X**) is overlapped by the supporting electrolyte decomposition wave. This concept is supported by the fact that unsubstituted stilbene yields two waves (at -2.18 and -2.56 V) in the same medium.

Linear concentration dependences were obtained for DPP at a DME over a range of (1-50) $\times$ 10^{-5} mol l^{-1} and for DPV at a HMDE in the range of 1×10^{-6}-5×10^{-4} mol l^{-1}. An attempt to use AdSV was not successful because of poor baseline at high negative potentials.

The developed methods can find application in the production and quality control of technical products, as well as in monitoring of this analyte in waste waters. In the latter case the DPP determination should be preceded by a preliminary TLC separation, as described in paper [69].

VIII

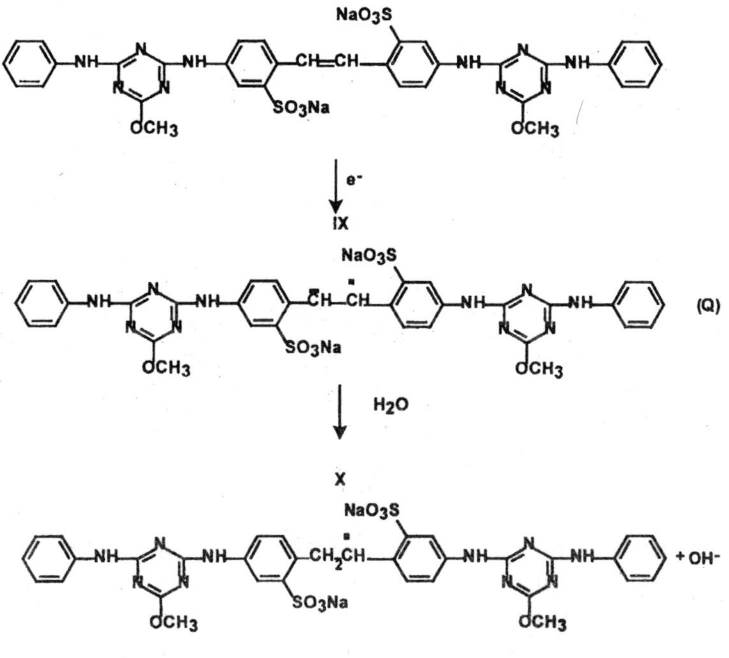

NaO_3S ... NH–(triazine, OCH_3)–NH–C$_6$H$_3$–CH=CH–C$_6$H$_3$–NH–(triazine, OCH_3)–NH–C$_6$H$_5$ ... SO_3Na

↓ e^-

IX

NaO_3S ... $\dot{C}H$–$\bar{C}H$ ... SO_3Na (Q)

↓ H_2O

X

NaO_3S ... $CH_2\dot{C}H$ ... SO_3Na + OH^-

7.6 Abbreviations

AC	acetate buffer
BASDA	4,4'-bis-[(4-phenylamino-6-methoxy-1,3,5-triazin-2-yl)amino] stilbene-2,2'-disulphonic acid
BR	Britton-Robinson buffer
CV	cyclic voltammetry
CPC	constant potential coulometry
DCV	direct current voltammetry
DPP	differential pulse polarography
DPV	differential pulse voltammetry
DME	classical dropping mercury electrode
DMF	dimethylformamide
$E_{1/2}$	half wave potential
E_p	peak potential
FS DPV	fast scan differential pulse voltammetry
h	mercury reservoir height
HMDE	hanging mercury drop electrode
HPLC-ED	high-performance liquid chromatography with electrochemical detection
IARC	International Agency for Research on Cancer
I_{lim}	limiting current
I_p	peak current
LOD	limit of determination
LSV	linear scan voltammetry
MeOH	methanol
MPTPC	7-[4-methyl-5-phenyl-2-(1,2,3-triazolyl)]-3-phenylcoumarin
RDE	rotating disk electrode
SCE	saturated calomel electrode
SMDE	static mercury drop electrode
TBABr	tetrabutylammonium bromide
TEABr	tetraethylammonium bromide
TLC	thin layer chromatography

7.7 References

1. Bond A. M.: *Modern Polarographic Methods in Analytical Chemistry*. M. Dekker, New York 1980.
2. Kissinger P. T., Heineman W. R. (Eds.): *Laboratory Techniques in Electroanalytical Chemistry*. M. Dekker, New York 1984.
3. Bard A. J., Faulkner L. R.: *Electrochemical Methods*. J. Wiley, New York 1980.

4. Kalvoda R., Volke J.: *Differential Pulse Polarography and Voltammetry.* In: *Instrumentation in Analytical Chemistry, Vol. 1* (Zýka J., Ed.). Ellis Horwood, Chichester 1991. P. 39.
5. Wang J.: *Electroanalytical Techniques in Clinical Chemistry and Laboratory Medicine.* VCH Publishers, Deerfield Beach 1988.
6. Wang J.: *Stripping Analysis.* VCH Publishers, Deerfield Beach 1985.
7. Wang J.: *Voltammetry Following Nonelectrolytic Preconcentration.* In: *Electroanalytical Chemistry, Vol. XVI* (Bard A. J., Ed.). M. Dekker, New York 1989. P. 1.
8. Kalvoda R., Kopanica M.: Pure Appl. Chem. *61*, 97 (1989).
9. Kalvoda R.: *Adsorptive Stripping Voltammetry.* In: *Instrumentation in Analytical Chemistry, Vol. 2* (Zýka J., Ed.). Ellis Horwood, Chichester 1994. P. 54.
10. Štulík K., Pacáková V.: *Electroanalytical Measurements in Flowing Liquids.* Ellis Horwood, Chichester 1987.
11. Kalvoda R., Parsons R. (Eds.): *Electrochemistry in Research and Development.* Plenum Press, New York 1985.
12. Smyth W. F. (Ed.): *Polarography of Molecules of Biological Significance.* Academic Press, London 1979.
13. Hart J. P.: *Electroanalysis of Biologically Important Compounds.* Ellis Horwood, Chichester 1990.
14. Bard A. J., Lund H. (Eds.): *Encyclopedia of Electrochemistry of the Elements. Vol. XI-XV: Organic Section.* M. Dekker, New York 1973-1984.
15. *IARC Monographs on the Evaluation of Carcinogenic Risk of Chemicals to Humans: Chemicals, Industrial Processes and Industries Associated with Cancer in Humans, Vol. 1-55.* IARC, Lyon 1970-94.
16. *IARC: Environmental Carcinogens - Selected Methods of Analysis. Vol. 1-13.* IARC, Lyon 1978-94.
17. Barek J., Hrnčíř R.: Collect. Czech. Chem. Commun. *51*, 2083 (1986).
18. Barek J., Kubíčková J., Mejstřík V., Petira O., Zima J.: Collect. Czech. Chem. Commun. *55*, 2904 (1990).
19. Barek J., Pastor T. J., Votavová S., Zima J.: Collect. Czech. Chem. Commun. *52*, 2149 (1987).
20. Barek J., Dřevínková D., Mejstřík V., Zima J.: Collect. Czech. Chem. Commun. *58*, 295 (1993).
21. Barek J., Hrnčíř R.: Microchem. J.: *36*, 172 (1987).
22. Barek J., Haladová-Bláhová H., Zima J.: Collect. Czech. Chem. Commun. *54*, 1549 (1989).
23. Barek J., Pastor T. J., Zima J.: Collect. Czech. Chem. Commun. *56*, 1210 (1991).
24. Barek J., Mejstřík V., Švagrová I., Zima J.: Collect. Czech. Chem. Commun. *56*, 2815 (1991).
25. Barek J., Pham Tuan Hai, Pacáková V., Štulík K., Zima J.: Fresenius J. Anal. Chem. *350*, 678 (1994).
26. Barek J., Švagrová I., Zima J.: Collect. Czech. Chem. Commun. *56*, 1434 (1991).
27. Barek J., Toubar S., Zima J.: Collect. Czech. Chem. Commun. *56*, 2073 (1991).
28. Barek J., Kubíčková J., Mejstřík V., Zima J.: Collect. Czech. Chem. Commun. *57*, 2263 (1992).

29. Barek J., Mejstřík V., Toubar S., Zima J.: Collect. Czech. Chem. Commun. *57*, 1230 (1992).
30. Ignjatovic L.M., Barek J., Zima J., Markovic D.: Anal. Chim. Acta *284*, 413 (1993).
31. Barek J., Dřevínková D., Zima J., Fogg A.: Collect. Czech. Chem. Commun. *57*, 2248 (1992).
32. Barek J., Dřevínková D., Zima J.: Collect. Czech. Chem. Commun. *58*, 2021 (1993).
33. Barek J., Fogg A. G.: Analyst (London) *117*,751 (1992).
34. Barek J., Malik G., Zima J.: Collect. Czech. Chem. Commun. *56*, 595 (1991).
35. Barek J., Matějka J., Zima J.: Collect. Czech. Chem. Commun. *59*, 294 (1994).
36. Barek J., Matějka J., Zima J.: Collect. Czech. Chem. Commun. *57*, 450 (1992).
37. Barek J., Berka A., Muller M., Zima J.: Collect. Czech. Chem. Commun. *50*, 2853 (1985).
38. Barek J., Zima J.: Collect. Czech. Chem. Commun. *54*, 361 (1989).
39. Barek J., Berka A., Muller M. ,Procházka M., Zima J.: Collect. Czech. Chem. Commun. *51*, 1604 (1986).
40. Barek J., Castegnaro M., Malaveille C., Brouet I., Zima J.: Microchem. J. *36*, 192 (1987).
41. Barek J.: Kelnar L.: Microchem.J. *33*, 239 (1986).
42. Barek J., Berka A., Zima J.: Talanta *32*, 987 (1985).
43. Barek J.: Microchem. J. *33*, 97 (1986).
44. Barek J., Berka A., Muller M.: Microchem. J. *33*, 102 (1986).
45. Castegnaro M., Malaveille Ch., Brouet I., Michelon J., Barek J.: Am. Ind. Hyg. Assoc. J.: *46*, 187 (1985).
46. Barek J., Pacáková V., Štulík K., Zima J.: Talanta *32*, 279 (1985).
47. Barek J., Berka A., Skokánková H.: Microchem. J. *29*, 350 (1984).
48. Castegnaro M., Barek J., Jacob J., Kirso U., Lafontaine M., S ansone E.B., Telling G.M., VuDuc T.: *Laboratory Decontamination and Destruction of Carcinogens in Laboratory Wastes: Some Polycyclic Heterocyclic Hydrocarbons*. IARC Scientific Publications No. 114. IARC, Lyon 1991.
49. Castegnaro M., Adams J., Armour J., Barek J., Benvenuto J., Confalonieri C., Goff U., Ludeman S., Reed D., Sansone E.B., Telling G.: *Laboratory Decontamination and Destruction of Carcinogens in Laboratory Wastes: Some Antineoplastic Agents.* IARC Scientific Publications No. 73. IARC, Lyon 1985.
50. Castegnaro M., Barek J., Dennis J., Ellen G., Klibanov M., Lafontaine M., Mitchum R., Van Roosmalen P., Sansone E.B., Sternson L.A., Vahl M.: *Laboratory Decontamination and Destruction of Carcinogens in Laboratory Wastes: Some Aromatic Amines*. IARC Scientific Publications No. 64. IARC, Lyon 1985.
51. Castegnaro M., Barek J., Frémy J. M., Lafontaine M., Miraglia M., Sansone E. B., Telling G. M.: *Laboratory Decontamination and Destruction of Carcinogens in Laboratory Wastes: Some Mycotoxins.* IARC Scientific Publications No. 113. IARC, Lyon 1991.
52. Burcinová A., Štulík K., Pacáková V.: J. Chromatogr. *389*, 397 (1987).
53. Rowe R. R., Smyth M. R.: *Electroanalysis of Agrochemicals*. In: *Polarography of Molecules of Biological Significance* (Smyth W. F., Ed.). Academic Press, New York 1979.

54. Volke J., Slamnik M.: *Polarography and Related Methods*. In: *Pesticide Analysis* (Das K. G., Ed.). M. Dekker, New York 1981. P. 175.
55. Barek J., Ebertová H., Mejstřík V., Zima J.: Collect. Czech. Chem. Commun. *59*, 1761 (1994).
56. Oláh B.: Thesis. Faculty of Science, Charles University, Prague 1992.
57. Hejhalová L.: Thesis. Faculty of Science, Charles University, Prague 1993.
58. Barek J., Tietzová B., Zima J.: Collect. Czech. Chem. Commun. *52*, 867 (1987).
59. Barek J., Balsiene J., Tietzová B., Zima J.: Collect. Czech. Chem. Commun. *53*, 921 (1988).
60. Barek J., Ghosh A., Zima J.: Collect. Czech. Chem. Commun. *54*, 1538 (1989).
61. Barek J., Civišová D.: Collect. Czech. Chem. Commun. *52*, 81 (1987).
62. Barek J., Balsiene J., Berka A., Hauserová I., Zima J.: Collect. Czech. Chem. Commun. *53*, 19 (1988).
63. Barek J., Švagrová-Hauserová I., Zima J.: Collect. Czech. Chem. Commun. *54*, 2105 (1989).
64. Barek J., Civišová D., Ghosh A., Zima J.: Collect. Czech. Chem. Commun. *55*, 379 (1990).
65. Barek J., Civišová D., Ghosh A., Zima J.: Collect. Czech. Chem. Commun. *55*, 1508 (1990).
66. Barek J., Hrnčíř R.: Collect. Czech. Chem. Commun. *59*, 309 (1994).
67. Barek J., Hrnčíř R., Moreira J. C.: Collect. Czech. Chem. Commun., in press.
68. Barek J., Hrnčíř R.: Collect. Czech. Chem. Commun. *59*, 1018 (1994).
69. Barek J., Hrnčíř R., Moreira J. C., Zima J.: Collect. Czech. Chem. Commun., in press.

8

ELECTROANALYSIS IN ENVIRONMENTAL CONTROL

Miloslav Kopanica and Ivana Šestáková

Abstract

Some aspects of possible application of several techniques of stripping voltammetry for the analysis of aquatic samples and biological material are demonstrated, together with a discussion of the properties and analytical applications of composite electrode materials containing graphite and silica gel.

Key Words

voltammetry, stripping voltammetry, composite electrodes, analysis of natural waters, determination of organic compounds, analysis of biological materials

8.1 Introduction

Sensitivity, possibility of automation and applicability to field measurements are the principal requirements on the methods of environmental analysis. From this point of view, electrochemical methods, especially voltammetry are very suitable, due to their high sensitivity and a wide range of applications, from the determination of metals to that of organic compounds. A high sensitivity and/or improvement in the signal-to-noise ratio can be attained by electrochemical preconcentration of the analyte and application of non-stationary measuring techniques in voltammetry [1]. Computer control of the voltammetric measurement then enables complete automation of the measurement itself and of the handling of the results of analyses [2].

8.2 Survey of methods

8.2.1 Determination of metals

For studying environmental problems and environmental control, simple and sufficiently sensitive methods for the determination of metals are necessary. For the determination of trace elements in aquatic systems (rivers, lakes, oceans, sediments, waste waters etc.), voltammetric techniques are very suitable due to their high sensitivity and relative simplicity and due to the fact that only a very simple pretreatment of the sample is required.

Voltammetry can further be applied to the analysis of samples containing metals in a range from ng per litre (free ocean water) to mg per litre (sediments). Another, often mentioned advantage of voltammetry, is the possibility of simultaneous determination of several metals - Cu, Pb, Cd and Zn; these metals of prime environmental concern can be determined by anodic stripping voltammetry simultaneously, with a high sensitivity.

Because of the high sensitivity of voltammetric methods and low amounts of metals in natural waters, the risk of contamination of the sample, the chemicals used, the vessels and the whole apparatus that comes in contact with the sample is high and must be prevented. The measurement itself has to be done within the shortest possible time after sampling. For prolonged storage it is best to rapidly freeze the filtered, acidified or untreated sample to -20 °C..

For trace element speciation in natural waters, special techniques of sample collection, filtration, storage etc. must be applied, to avoid contamination or losses of the trace metals. On the other hand, voltammetric techniques require only a simple sample handling and thus decrease the contamination risk and further enable discrimination of labile/inert complexes in natural waters and distinguishing different valency states of elements.

If biological or geochemical cycles of elements in natural waters are studied, measurement of instantaneous concentration changes in real time and within required space is necessary. *In situ* measurement is the only solution to this problem; this type of measurement also removes the problems connected with sample pretreatment, artifacts due to chemical changes of the probe (removal of oxygen) and problems due to adsorption and contamination. Voltammetry is very well suited for such type of measurement, submersion voltammetric probes equipped with flow-through cells and mercury film electrodes have been developed and used for the determination of Cu, Pb, Cd and Zn.

8.2.2 Determination of organic compounds

Among the enormous number of organic substances, there are many which can be determined voltammetrically. Three kinds of current can be exploited in the analysis of organic compounds:

1. Diffusion controlled reduction or oxidation currents
2. Catalytic currents (e.g., sulphur containing proteins in ammoniacal cobalt solutions)
3. Adsorption currents (tensammetric waves)

An extreme increase in the sensitivity is attained by preconcentrating the species to be determined through adsorption (adsorptive stripping) or anodically, as poorly soluble salts or complexes with mercury ions, followed by cathodic stripping.

Organic compounds in the environment are degraded or metabolized in living organisms and thus their stability, fate and metabolites produced should be known for successful analysis. Separation methods are usually necessary, which include, e.g., selective extraction, chromatography and electrophoresis, or the use of preconcentration columns.

8.3 Results and Discussion

8.3.1 Water

8.3.1.1 DETERMINATION WITH Hg ELECTRODE

Stripping voltammetry - electrochemical enrichment

Differential-pulse anodic-stripping voltammetry (DPASV) with a hanging mercury drop electrode (HMDE) or a mercury film electrode (MFE) on a glassy carbon support is the most common method for the determination of trace elements in natural waters and is applicable for the determination of amalgam-forming metals. If the formation of an amalgam is kinetically complicated, as in the case of manganese deposition, it is preferable to apply the method of electrochemical enrichment [3, 4]: Manganese is first deposited on the surface of a HMDE as in ASV and, after a short oxidation interval at more positive potential, the re-formed manganese(II) ions are determined by differential pulse (DP) cathodic polarization scan [5]. Under these conditions, the solution in the vicinity of the electrode is electrochemically enriched in the manganese(II) ions and, when the duration of the oxidation interval is properly chosen, the sensitivity of the method is practically equal to that of DPASV.

Non-faradayic preconcentration of analyte

In voltammetric determination of trace metals, the analyte is preconcentrated on the electrode surface by electrolysis. If this is impossible, it can be accumulated on the electrode surface by adsorption or by a reaction with the material of the working electrode. The determination of thiourea and related compounds is a typical example, the accumulation of the analyte being based on the reaction between mercury(II) ions, formed by anodic dissolution of the mercury electrode, and the analyte [6].

Similarly, the determination of traces of selenium by cathodic stripping voltammetry with a HMDE involves the formation of mercury(II) selenide at the deposition potential close to 0 V, followed by the reduction of mercury(II) selenide during the cathodic polarization scan. For the determination of very low amounts of selenium, in the range below 0.8 ng per ml, it is preferable to apply adsorptive stripping voltammetry, where selenium is preconcentrated on the HMDE surface as the selenium-3,3'-diaminobenzidine complex [7].

Adsorptive stripping voltammetry of uranium in the form of its complex with catechol [8] has been found very successful for the determination of this element in natural waters and modified for routine determination of uranium in sea water.

Measurement of catalytic currents belongs among extremely sensitive voltammetric methods, applicable to environmental analysis. An example is the DP voltammetric determination of molybdenum in a nitrate medium, in the presence of 8-hydroxyquinoline [9].

The compounds which cannot be electrochemically reduced or oxidized at a suitable electrode material but which are adsorbed at the electrode surface, can be determined voltammetrically with a sufficient sensitivity. Such a compound is trichlorobiphenyl (TCB) and other chlorinated biphenyls. The examination of the electrocapillary curve of mercury measured in a TCB solution of pH 6.8 containing 20 % methanol revealed that TCB is adsorbed at the electrode surface at a potential of -0.40 V(*vs.* Ag/AgCl) and is desorbed -1.05 V. On this basis, a very sensitive determination, analogous to cathodic stripping voltammetry, has been developed: The analyte is preconcentrated by adsorption at a HMDE at -0.40 V and stripped during a cathodic potential scan with a peak at -1.05 V [10]. The detection limit amounts to 0.003 mg l^{-1}; the interference from DDT can be eliminated chemically by treatment with dilute sulphuric acid at 80 °C for 15 min.

8.3.1.2 DETERMINATION WITH CARBON COMPOSITE ELECTRODES

The mercury film electrode (MFE), especially in the form of a rotating disk electrode (RDE), is generally accepted as the best working electrode for the determination of traces of amalgam-forming metals. The MFE is commonly prepared by depositing mercury on the surface of a glassy carbon electrode. During a study of carbon composite electrodes [11, 12], attention was paid to the use of these electrodes in stripping voltammetry. The studied electrodes, which are fabricated by mixing graphite powder, ceresine wax and silica gel at an elevated temperature, belong to the type of electrodes with a random distribution of conductor particles within the insulator matrix. A typical current-potential response of these electrodes is a sigmoidal curve and the dependence of the limiting current on the potential scan rate exhibits a region within which the limiting current is practically independent of the scan rate. These findings confirm that this type of composite electrode behaves as an ultramicroelectrode array. As expected, a signal is obtained even when the measured solution contains no electrolyte. The voltammetric response of a composite electrode containing 10 % of silica gel with a specific surface area of 640 m^2 g^{-1} in a ferrocyanide solution was sigmoidal and the slope of the log $(i_d - i)/i$ *vs.* E dependence is 44 mV in 0.01M KNO_3 and 46 mV in the absence of the electrolyte. Evidently, composite electrodes of this type may be used for *in situ* voltammetric measurements.

These electrodes have also been tested as working electrodes in anodic stripping voltammetry (ASV). As follows from the results obtained by ASV of silver in ammoniacal media, stirring during the pre-electrolysis has practically no effect on the height of the appropriate anodic peak, if the electrolysis time is short. Composite electrodes of the described type have further been used as supports for the MFE preparation, by *in situ* deposition of mercury. Similar to voltammetric measurements with carbon composite electrodes, the measurement with such a modified MFE exhibits a smaller effect of stirring during the preelectrolysis on the stripping peak current values. The background current also decreases, in proportion to the specific surface area of the silica gel used.

The removal of dissolved oxygen from the analyzed solutions is necessary in work with macroelectrodes due to the reduction of oxygen. For field measurements, a portable device for oxygen removal has been developed employing a stream of carbon dioxide or nitrogen oxide from small household pressure bottles [13].

If dissolved oxygen is not removed from the analyzed solution, acceptable results can be obtained using high polarization rates (1 to 5 V s^{-1}) during the anodic dissolution step [14], as the oxidation of the amalgam is then suppressed.

8.3.2 Biological materials

8.3.2.1 METALS

To determine very low concentrations of metals in various biological materials, voltammetric methods involving accumulation of the measured species at the working electrode are used, namely, anodic stripping voltammetry (ASV), cathodic stripping voltammetry (CSV) and adsorptive stripping voltammetry (AdSV), usually in the differential pulse mode. With the exception of some liquid samples (urine, plasma), biological samples must be solubilized and mineralized prior to the measuring step. There are several possibilities how to do this, such as classical dry ashing (performed in air, at the atmospheric pressure), pressurized wet digestion, microwave digestion etc. The main reason for this treatment is the necessity to transform the metal present in biological tissues in several chemical forms into a single chemical species with a uniform response and to overcome the heterogeneity of the biological material and the interferences from the organic matrix. It has been shown that an incompletely digested organic matrix can influence the results of anodic stripping voltammetric measurements of Pb, Cd and Cu, especially when nitric acid is used and organic nitro-compounds are formed during the first stages of digestion [15, 16]. Using adsorptive stripping voltammetry, the formation and degradation of these compounds can be followed and the digestion procedure optimized [17]. Such an optimized procedure was used for dry ashing in superoxidizing atmosphere (mixture of $O_2 + O_3 + NO_x$), when a reference material of animal origin was analyzed independently by atomic absorption spectroscopy (AAS) and anodic stripping voltammetry, with computer-controlled ECO-TRIBO Polarograph [18]. The digests were halved; one half was analyzed using electrothermal AAS (ETAAS) for Cd and Pb and using flame AAS for Cu. The other half was analyzed by DPAS. A statistical evaluation of the results showed a good correlation between the two methods (the correlation coefficients r equal to 0.997 for Cd, 0.993 for Cu, and 0.991 for Pb) and the results of the two methods were in agreement with the certified values.

Whereas some metals are essential or exhibit toxic effects only at higher concentrations, mercury, lead and cadmium are toxic to living organisms at any level. Therefore, all compounds containing these metals are highly interesting. Metallothioneins (MT) are a unique class of low-molecular weight proteins, characterized by a high content of cysteine and a capacity to bind a broad spectrum of metals, with the highest affinity towards mercury and cadmium. Metallothioneins have been found in various living organisms ranging from fungi to the mammals; they probably protect the organisms against overloading with toxic metal ions and maintain a homeostasis of the essential metal ions, such as zinc and copper [19]. Fundamental studies have been performed on Cd- and Zn-containing rabbit-liver metallothioneins, using spectroscopic and NMR techniques [20, 21]. Despite using Brdička's reaction to determine the total amount of sulphur-containing proteins, electrochemistry has not yet obtained unambiguous results. A comparison of the results obtained with mercury and carbon electrodes shows that the behavior of CdMT and Cd,ZnMT is complicated by the formation of mercury compounds [22]. To prevent this, the influence of surfactants, such as TPO and TPPO are studied [23], obtaining for Cd,ZnMT a linear dependence of the reduction peak on the metallothionein concentration over a range from 8.9×10^{-10} mol dm^{-3} to 5×10^{-5} mol dm^{-3}.

8.3.2.2 ORGANIC COMPOUNDS

Voltammetry after adsorptive accumulation at an HMDE has been used for determination of cyadox and its metabolites in plasma. The total concentrations of the original compounds and three possible metabolites has been measured on the basis of the two-electron reduction of the quinoxaline ring, common to all the four compounds involved. Two separation procedures have been successful: gel chromatography (Sephadex G 25) or selective extraction followed by preconcentration on a Sep-Pak micro-column. The sensitivity of adsorptive stripping voltammetric determination is in a range of ng ml^{-1} and the results are in agreement with a radiometric determination [24].

A nitrofurane derivative, furazolidon, 3-(5-nitrofurfuryliden amino)-2-oxazolidon, yields a voltammetric peak due to the four-electron reduction of nitrogroup in acid media at an HMDE in analyses of samples of liver, kidney and meat. Acetone extraction of the freeze-dried material and thin-layer chromatography have been used for sample preparation. The limit of determination was 0.03-0.007 μg g^{-1} in dependence on the kind of the tissue [25].

The catalytic wave of the hydrogen reduction in an ammoniacal solution of cobalt salts (Brdička DPP assay) is used for determination of sulphur-containing proteins [26], with limits of determination of the order of 10^{-12} mol dm^{-3}. In this case, thermostability of polymers with high molecular weights is used for separation from other interfering compounds. Application of this method for analysis of sulphur-containing polymers from plant materials is currently studied [27].

8.4 Conclusions

Voltammetry, especially all the types of stripping voltammetry, is among the methods which are perfectly suitable for the determination and speciation of trace metals in aquatic systems. The development of microelectrodes and new electrode materials leads to the use of voltammetry for *in situ* measurements, i.e., in the presence of dissolved oxygen and in the absence of supporting electrolyte. The determination of trace metals in biological materials by anodic stripping voltammetry is reliable, provided that an efficient method is employed for mineralization of the analyzed material. In voltammetric determinations of organic substances in biological materials, differences in the electroactivity of the analyzed and interfering compounds can be exploited in simplified separation procedures.

8.5 References

1. Kalvoda R., Ed.,: *Electroanalytical Methods in Chemical and Environmental Analysis*: Plenum Press,N.York,1987.
2. Novotný L. et al.: 4th European Conference on Electroanalysis, Noordwijkerhaut, 1992.
3. Yarnitsky Ch., Ariel M.: J. Electroanal. Chem. *10*, 110 (1965).
4. Kopanica M. , Stará V.: J. Electroanal. Chem. *127*, 255 (1981).
5. Stará V., Kopanica M.: Electroanalysis *5*, 595 (1993).
6. Kopanica M.: The Science of the Total Environment *37*, 83 (1984). P. 1.
7. Stará V., Kopanica M.: Anal. Chim. Acta *208*, 231 (1988).
8. Lam N. K., Kalvoda R., Kopanica M.: Anal. Chim. Acta *154*, 79 (1983).
9. Navrátilová Z., Kopanica M.: Anal. Chim. Acta *244*, 193 (1991).
10. Lam N. K., Kopanica M.: Anal. Chim. Acta *161*, 315 (1984).
11. Stará V., Kopanica M.: Electroanalysis *1*, 251 (1989).
12. Kopanica M., Stará V.: Electroanalysis *3*, 13 (1991).
13. Langmaier J., Opekar F., Šestáková I.: Chem. Listy *85*, 102 (1991).
14. Kopanica M., Stará V.: Electroanalysis *3*, 925 (1991).
15. Pratt K., Kingston H. M., MacCrehan W. A., Koch W. F.: Anal. Chem. *60*, 2024 (1988).
16. Mader P., Musil J., Čurdová E., Korečková J., Cibulka J.: Chem. Listy *81*, 1190 (1987).
17. Mader P., Miholová D., Šestáková I., Száková J.: *Proc. J. Heyrovský Centennial Congress on Polarography, vol. I*, Prague 1990. P. Tue86.
18. Šestáková I., Miholová D., Slámová A., Mader P., Száková J.: Electroanalysis *6*, 1057 (1994).
19. Stillman M. J., Shaw III, C. F., Suzuki K. T., Eds.: *Metallothioneins*. VHS Publishers, New York,1992.
20. Good M., Hollenstein R., Vašák M.: Eur. J. Biochem. *197*, 655 (1991).
21. Robbins A. H., McRee D. E., Williamson M., Collett S. A., Xuong N. H. , Furey W. F., Wang B. C., Stout C. D.: J. Mol. Biol. *221*,1269 (1991).
22. Šestáková I., Miholová D., Vodičková H., Mader P.: Electroanalysis *7*, 237 (1995).
23. Fedurco M., Šestáková I.: Bioelectrochem. Bioenergetics *36* (1995), in print.
24. Šestáková I., Kopanica M.: Talanta *35*, 816 (1988)
25. Šestáková I., Dušková S.: Biopharm. *1*, 199 (1991).
26. Hogstrand Ch., Haux C.: Anal. Biochem. *200*, 388 (1992).
27. Šestáková I., Vodičková H., Mader P., Miholová D.: in preparation.

9

ELECTROCHEMISTRY OF BIOLOGICALLY ACTIVE SUBSTANCES IN NON-AQUEOUS MEDIUM

Ivan Němec, Naďa Zimová

Abstract

Electrochemical methods in acetonitrile medium have been employed for studying electron-transfer processes of biologically important molecules. Mechanism and kinetics of electrode processes of triazines, phenothiazines, thiobenzamides, thiobenzanilides were solved, as the model for metabolic oxidation. Attention was focused on the electron-donor behaviour of the studied substances, substituent structure and position.

Key Words

voltammetry, oxidation, radicals, acetonitrile, triazines, phenothiazines, thiobenzamides, thiobenzanilides

9.1 Introduction

From the point of view of physico-chemical interactions in the environment it is highly important to investigate electron-transfer processes of biologically important molecules; voltammetric methods are useful in model studies of receptor and enzyme interactions with these substances. Particularly important are heterocyclic systems containing nitrogen, as well as other nitrogen-containing substances. They are involved in many reactions of living organisms. Studying biological active substances *in vivo* is difficult because of the complexity of living organisms. Simplified model systems are very useful because they can be defined well and yield useful information about investigated substances and their reaction mechanisms. These data, together with thermodynamic parameters, provide information important for understanding electron-transfer processes of biologically important substances in human body and the environment [1].

Our attention has been focused on electron-transfer processes and on the study of radical ions. Electron-transfer reactions are important from the point of view of living organisms, radical ions are significant in cancerogenesis, ageing, metabolism and other biologically important processes; they are also found in the environment. Other important reactions are solar energy conversion, photosensitivity of skin and eye tissues in the treatment of various diseases, environmental influences on metabolism, influence of pesticide traces in food, etc.

The main reason for using non-aqueous media is mainly a poor solubility or insolubility in water of many substances contained in ecosystems and human body. Other important advantages of non-aqueous solvents have been discussed elsewhere [2]. The simplicity in non-aqueous solvents and a limited effect of homogeneous chemical reactions help to solve some aspects of more complex processes occurring in water, the environment and human organism. It is possible to compare electrochemical oxidations in non-aqueous media with enzymatic or chemical oxidations. Most of the substances studied by us were water insoluble. Therefore acetonitrile (AN) medium was used. The reaction intermediates and products were then more stable and hydrolysis, as well as other homogeneous reactions, were suppressed.

Electrochemistry is a very useful technique in these model biological and pharmacological studies; on the contrary, biology offers various approaches to electrochemistry, e.g. biosensors. Electrochemistry permits generation and identification of reaction intermediates, particularly radical species. Radicals have been subject of much attention in recent electroorganic studies, as they can play an important role in metabolism of many substances.

Our studies of the electrochemical behaviour of certain groups of biologically active substances employed voltammetric techniques in AN medium, particularly cyclic voltammetry, CV, (scan rates from 0.5-17 V s^{-1} to 70 V s^{-1}), DC voltammetry with a rotating disc electrode, RDE, (1226-2100 rpm), controlled potential coulometry and constant-potential electrolysis. A solution of 0.1M sodium perchlorate was used as the supporting electrolyte. The working electrodes were a rotating platinum disc (RDE), a stationary platinum disc (CV) or a platinum wire (constant potential coulometry and electrolysis). A silver electrode immersed in 0.01M silver nitrate in AN was used as the reference electrode.

An acid AN medium was also used to study proton influence on the electrochemical behaviour of the studied derivatives; a combined effect of water and acid was investigated, as $HClO_4$ was a 70 % aqueous solution.

9.2 Results and Discussion

9.2.1 Triazine herbicides

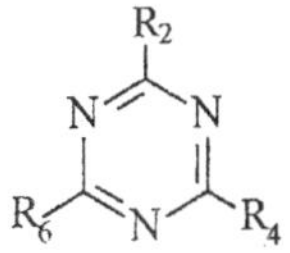

The studied substances were s-triazine derivatives, substituted in 2, 4 and 6 positions of the heterocyclic skeleton. The electrochemical processes of triazine derivatives are still insufficiently described in the literature. s-Triazines belong to the most important selective herbicides and they are highly persistent. They are known by a high acute toxicity, and undergo hydrolytic and oxidation processes in soil. Toxic hydrolytic and oxidation products should thus be monitored from the point of view of veterinary practice and agriculture. In an anhydrous AN medium, the oxidation mechanism is simplified and can readily be clarified. The electrochemical behaviour of s-triazine derivatives was investigated [3] using DC voltammetry at an RDE and cyclic voltammetry. The DC and CV curves in 0.1M $NaClO_4$ in AN exhibited one irreversible anodic wave or peak within a potential scan rate of 0.5-0.8 V s^{-1} for all the studied derivatives. The calibration curves were linear within a concentration range of 5×10^{-5} to 7×10^{-4} mol dm^{-3}. The peak potentials and the half-wave potential values were independent of the concentration of the studied derivatives. The CV criteria proved irreversibility of the processes; no cathodic peak was observed at given scan rates (***Fig. 1***).

Chloro-s-triazines were oxidized at higher potential values than the methoxy- and thiomethylderivatives (***Table 1***). Only slightly different half-wave potentials were observed for the methoxy- and thiomethylderivatives.

In the series of chloro-s-triazines, the half-wave potential depended on the number of alkyl groups per amino group (in position 4 and 6). Replacement of hydrogen atoms by alkyl groups led to a considerable decrease in the half-wave potential values (simazine *vs.* triethazine, atrazine *vs.* ipazine).

On the other hand, the length of the carbon chain had no perceptible effect on the half-wave potentials. The s-triazine itself was not oxidized; the exocyclic aminogroups in the 4 and 6 positions acted as electron donors. The presence of an electronegative atom, chlorine, decreased the electron density on the nitrogen atom and increased the half-wave potential value. s-Triazine derivatives with methoxy or thiomethyl group in the position 2 and derivatives containing tert. amino groups were readily oxidized in anhydrous AN medium; they had the lowest half-wave and peak potentials.

***Tab. 1:* The half-wave potentials $E_{1/2}$ (V) of some s-triazines**

Common name	R_2	R_4	R_6	$E_{1/2}$ [V]	pK_B (ref. [4])
s-Triazine	H	H	H	not oxidized	-
Simazine	Cl	NHEt	NHEt	1.77	12.35
Triethazine	Cl	NHEt	$N(Et)_2$	1.57	-
Atrazine	Cl	NHEt	NHiPr	1.82	12.32
Ipazine	Cl	$N(Et)_2$	NHiPr	1.57	-
Propazine	Cl	NHiPr	NHiPr	1.75	12.15
Tertbutylazine	Cl	NHEt	NHtBu	1.81	12.06
Simetone	OCH_3	NHEt	NHEt	1.59	9.83
Prometone	OCH_3	NHiPr	NHiPr	1.57	9.72
Terbutone	OCH_3	NHtBu	NHEt	1.59	9.54
Simetryne	SCH_3	NHEt	NHEt	1.59	9.96
Ametryne	SCH_3	NHEt	NHiPr	1.57	9.90
Prometryne	SCH_3	NHiPr	NHiPr	1.63	9.95
Terbutryne	SCH_3	NHEt	NHtBu	1.58	9.62
Desmetryne	SCH_3	NHMe	NHiPr	1.58	-

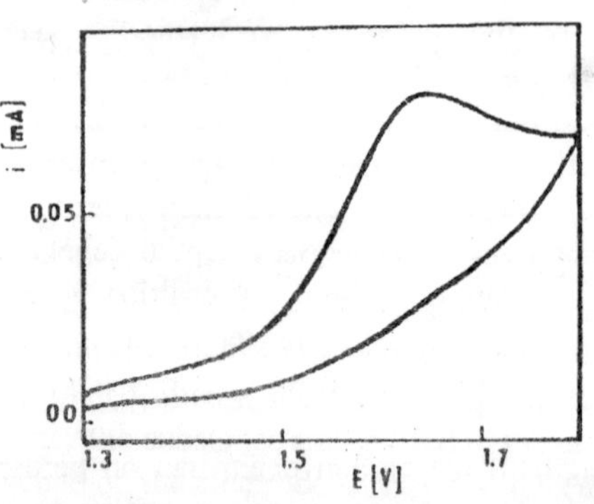

Fig. 1: **Cyclic voltammogram of the anodic oxidation of triethazine ($c = 8.3 \times 10^{-4}$ mol dm^{-3}) in an AN solution**
0.1M $NaClO_4$ at 20 °C, $A = 0.126$ cm^2, $v = 0.71$ V s^{-1}

The half-wave potential values correlated well with the pK_B values obtained in an aqueous solution. Substances with the highest pK_B (the lowest basicity) had the highest half-wave and peak potentials (***Table 1***).

An acid ($HClO_4$) AN medium caused a decrease in the anodic waves or peaks for all the studied derivatives, probably due to protonation of both the electron donor amino groups; their oxidation was thus impossible. The oxidation apparently did not occur at the s-triazine cycle; the triazine cycle was not decomposed. It is known from environmental studies [5], that triazines are degraded in soil through hydrolysis, oxidation, the effect of solar radiation and soil microorganisms. Dealkylation in position 4 and 6 and hydrolysis of the substituent in position 2 primarily occur. The s-triazine cycle is stable in soil for years. Our studies confirmed their high persistence, alkyl substituent influence on the oxidation processes and dependence between the basicity of the studied derivatives and their resistance to oxidation. Our studies, carried out in an anhydrous AN medium, are, however, also valid for aqueous solutions and the environment.

9.2.2 Phenothiazine derivatives

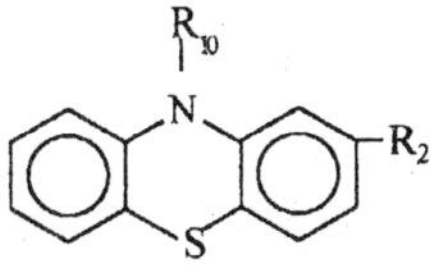

R_{10}

N

R_2

S

Phenothiazines are important in several ways for the human environment. Phenothiazine and some of its quinoid derivatives were first synthesised about 100 years ago and were important in dye chemistry. They were also studied in connection with the construction of galvanic cells, solar energy conversion and as photosensitizers. However, the most studies have been carried out because of their medical applications [6]. Phenothiazines exhibited a wide spectrum of biological and pharmacological activity. Despite a great size of clinical, medical and pharmacological literature concerned with phenothiazines, the mechanism of their biological action has not been fully resolved. Phenothiazines are good electron donors and a study of their electron-transfer processes and the understanding of the factors determining electron-transfer rates are of considerable importance and should help in the understanding of the mechanism of their action in living organisms.

Phenothiazine and some of its C-substituted derivatives have antihelmintic activity. The toxic antihelmintic effect probably involves interference of the phenothiazine radical cation with the redox couple in parasite; the redox potentials are probably important in the mechanism of these interactions [7]. *N*-substituted phenothiazine derivatives have antihistaminic activity. *N*- and 2-substituted derivatives are used for the treatment of mental illnesses [6]. Some of them also have antibacterial, anticancer, antituberculotic and antivirus activities [1, 8].

The environment can influence phenothiazine side effects in living organisms; frequent use of phenothiazine pharmaceuticals causes photosensitivity of skin and eye tissues, probably due to environmental effects and the melanine interactions with the metabolic phenothiazine products [6]. These interactions can be studied by electrochemistry, as they depend on the electron-donor properties of the reactants.

The mechanism of the electron-transfer reactions of phenothiazines is complex, particularly in aqueous medium. The basic electrochemical oxidation of phenothiazine and its derivatives was studied by Billon and co-workers [9, 10] and lately re-examined by Cauquis and co-workers [11, 12]. Since then, the importance of phenothiazines has increased.

Our attention was focused on the electron-donor behaviour of phenothiazines, the influence of the substituent position and the structure on the electrochemical behaviour and particularly on the radical cation formation in an AN medium. It was suggested that the radical cation may be the active pharmacological entity [6].

Phenothiazine and 26 of its derivatives were investigated in an AN medium [13-18]. The derivatives were simple 2-substituted (chlorine, methoxy, acetyl, trifluoromethyl) and 10-substituted (methyl, ethyl, propyl, butyl, isopropyl, isobutyl, and benzyl) phenothiazines and phenothiazine pharmaceuticals, substituted either in position 10 or in positions 2 and 10 (15 derivatives: promazine, diethazine, promethazine, profenamine, alimepromazine, levomepromazine, chlorpromazine, prochlorperazine, perphenazine, trifluoperazine, fluphenazine, periciazine, cyamepromazine, thioproperazine, levomepromazine, thioridazine). The electrochemical oxidation was studied by DC and cyclic voltammetry. Potentiostatic coulometry was used to determine the number of electrons exchanged. An acid AN medium was also used. The acid concentration ($HClO_4$) was in a range of 0.02 to 1.5 mol dm^{-3} (the water content was from 0.06 to 3.8 %; $HClO_4$ was a 70 % aqueous solution, thus a combined effect of water and acid was investigated).

9.2.2.1 ELECTROCHEMISTRY IN NON-AQUEOUS MEDIUM

Only the simple phenothiazine derivatives were studied in a non-aqueous AN medium [14, 16], as the electrochemical behaviour of phenothiazine pharmaceuticals was complicated. These derivatives are salts of various acids of a different strength; it is thus impossible to compare the voltammetric results without adding strong acid ($HClO_4$). The 2-substituted derivatives were oxidized in one step, two electrons were exchanged. The 10-substituted derivatives were oxidized in two steps; the first anodic step was reversible, while the second one was irreversible or quasireversible (at high scan rates, v = 50-270 V s^{-1}). The electrooxidaton potentials depended on the substituent position and structure. Generally, the simple 10-substituted derivatives are oxidized at higher potential values.

9.2.2.2 ELECTROCHEMISTRY IN ACIDIC ACETONITRILE MEDIUM

In an acidic AN medium, phenothiazine and its 2-substituted derivatives were oxidized in two steps [14]. Derivatives substituted on the heterocyclic nitrogen atom were oxidized in three or four steps (*Fig. 2a*) [15,16].

Radical cation formation

All the studied derivatives were oxidized to the corresponding radical cations, as was previously described in the literature (*Fig. 2b*) [9, 10]. One-electron exchange was confirmed by potentiostatic coulometry and by semilogarithmic analysis of DC and CV curves. The reversibility was studied by CV; cyclic voltammograms had quasireversible features. The degree of reversibility and radical cation stability increased with increasing acid concentration.

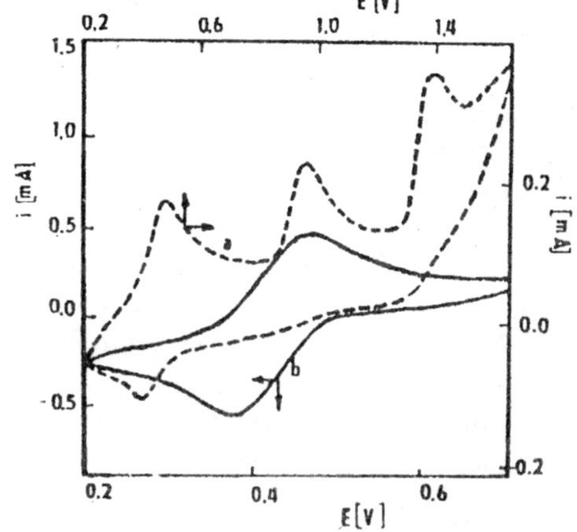

Fig. 2: **Cyclic voltammogram of levomepromazine ($c = 7 \times 10^{-4}$ mol dm^{-3}) in an AN solution of 0.1M $NaClO_4$ and 0.5M $HClO_4$,**
t = 20 °C, A = 0.288 cm^2 ; a) v = 0.9 V s^{-1} , b) v = 16 V s^{-1}

An exceptional behaviour was observed with simple 10-substituted derivatives [16]. Depending on a combined effect of the acid and water, the potential scan rate and substituent structure, inhibition of the first anodic step was observed. This was caused by protonation of the heterocyclic nitrogen atom. The protonated form was not electrochemically oxidizable; its dissociation was required prior to the charge transfer reaction ($K_A = 1.2 \times 10^{-3}$). An analogous behaviour was observed for phenothiazine dyes. The limiting diffusion currents and peak currents varied as a function of the acid concentration. 10-Isopropyl phenothiazine was oxidized at the highest potential values; the inhibition of its radical cation formation was observed at the lowest acid concentration, 0.02M $HClO_4$, $v = 1.25$ V s^{-1} (inhibition for 10-methyl derivative: 1.0M $HClO_4$ in AN, $v = 1.3$ V s^{-1}; for benzyl and propyl derivative, 0.5M $HClO_4$ in AN, $v = 50$ V s^{1}; for isobutyl derivative 1 to 1.5M $HClO_4$, $v = 0.5$ V s^{-1}). This effect was not observed in voltammetry at an RDE and during the cathodic step. The peak potentials and half-wave potentials linearly increased with increasing acid concentration, as did the limiting diffusion currents or peak currents. This behaviour indicated a high sensitivity toward preceding homogeneous chemical reactions. A similar behaviour was observed with some phenothiazine pharmaceuticals. However, the acid influence was not so strong and complete inhibition of the first anodic step did not occur.

On the contrary, the electrooxidation potentials for the radical cation formation were constant or decreased with increasing acid concentration with the derivatives substituted in position 2 and phenothiazine pharmaceuticals [14, 15]. The exceptions were promazine and thioridazine (their half-wave and peak potentials increased with increasing acid concentration). An increase in the potential scan rate from 0.5 to 17 V s^{-1} resulted in an improved peak potential separation, increased peak width and higher currents. The difference between the first two oxidation steps also increased, while the current function remained almost constant. The CV parameters depended on the acid concentration, except for the peak current ratios. They were close to unity, independent of the acid concentration and potential scan rate (for scan rate $v = 0.5\text{-}17$ V s^{-1}). At a high acid concentration (0.1 to 1.5M $HClO_4$ in AN), strong participation of the acid and water in the complex reaction mechanism must be considered.

The diffusion coefficients for the radical cation formation were determined from the Levich equation, using voltammetry at an RDE. They were in agreement with the values obtained by cyclic voltammetry. The diffusion coefficients decreased with increasing acid concentration in the AN medium (this was in agreement with the observed decrease in the limiting diffusion and peak currents), as the kinematic viscosity of the reaction medium increased. A linear relationship was found between the estimated diffusion coefficients and the reciprocal kinematic viscosity. The diffusion coefficients were in a range of 10^{-6}-10^{-5} $cm^2 s^{-1}$ [18]. It is interesting that the diffusion coefficients primarily depended on the combined effect of perchloric acid and water in the AN medium, whereas the substituents exerted no important influence.

Cyclic voltammetry was used to calculate the standard heterogeneous charge-transfer rate constants for the radical cation formation. The peak separation was used to determine the heterogeneous rate constants using the Nicholson-Shain approach [19]. The electrode reaction was accelerated and the heterogeneous rate constants linearly increased for all the derivatives, with the perchloric acid concentration in AN (0.02-0.5M). At higher acid concentrations (0.5-1.5M $HClO_4$ in AN) a linear increase in the heterogeneous rate constants was only found for 2-substituted derivatives. A decrease was observed for derivatives substituted at the heterocyclic nitrogen atom (0.5M $HClO_4$ in AN for simple 10-substituted derivatives and 0.9M $HClO_4$ in AN for phenothiazine pharmaceuticals). The heterogeneous rate constants were of the order of 10^{-2}-10^{-1} cm s^{-1} [18]. The structure and position of the substituent had no influence on the electrode kinetics at low acid concentrations (0.02-0.5M $HClO_4$ in AN). The position of the substituent affected the heterogeneous rate constants when the acid concentration (and thus also that of water) increased (0.5-1.5M $HClO_4$).

Along with the basic voltammetric study, the electron donor behaviour of the substances was also investigated [14, 15]. A substitution at the heterocyclic nitrogen atom caused deterioration in the electron donor properties that is probably due to a change in the molecular geometry. The deplanarization of the system must be connected with a decreased conjugation of the molecule leading to an increase in both the ionization potential and the electrooxidation potential. Steric reasons are probably the main cause of an increase in the peak and half-wave potentials for derivatives substituted at the heterocyclic nitrogen atom of the phenothiazine nucleus. Not only the substituent position, but also the structure of the substituent in position 10 played an important role considering the electron properties. The oxidation was easier with straight-carbon-chain substituents than with branched ones and branching on the first carbon atom of the substituent chain caused a greater impediment than that on the second carbon atom [16]. Introduction of an aromatic ring to the second carbon atom caused an effect analogous to that of branching. A similar behaviour was observed for phenothiazine pharmaceuticals [15]. Substitution at the propylene or ethylene side chain by a nitrogen containing group (pheno-thiazine pharmaceuticals) caused a further increase in the half-wave and peak potential values (10-propylphenothiazine *vs.* 10-dimethylaminopropyl-phenothiazine, i.e. promazine; 10-ethylphenothiazine *vs.* 10-diethylamino-ethylphenothiazine, i.e. diethazine). A branched side chain caused an increase in the half-wave and peak potentials, similar to the presence of an aromatic ring in the side substituent (diethazine-profenamine, periciazine-cyamepromazine; 10-isopropylphenothiazine, 10-isobutylphenothiazine, 10-benzylphenothiazine). The effect of branching is increased by a strong non-bonding interaction with alkyl groups at the side chain nitrogen atom; inductive effects are also important (prophenamine *vs.* promethazine).

Several exceptions were observed. The half-wave potentials for the thioridazine and promazine radical cation formation increased with increasing acid concentration. This behaviour was similar to that of simple 10-substituted derivatives. Protonation of the side chain nitrogen atom is probable [15,16].

In the case of phenothiazine pharmaceuticals, a trimethylene group localized between the nitrogen atom of the phenothiazine skeleton and the nitrogen atom of the side chain effectively isolated the latter nitrogen from the phenothiazine nucleus. The changes in the structure at more distant positions have virtually no effect on the half-wave potential values of the radical cation formation (chlorpromazine *vs.* perphenazine *vs.* prochlorper-azine, fluphenazine *vs.* trifluoperazine, cyamepromazine *vs.* periciazine[15].

The half-wave potential and peak potential values for the derivatives substituted in position 2 and in positions 2 and 10 followed the σ_p Hammett constants. The stability of the radical cation and the aromatic character of molecules with an electron donating substituent (e.g. $-OCH_3$, $-SCH_3$) increased and the electrooxidation was easier. Electrooxidation of derivatives with electron withdrawing substituents ($-CF_3$, $-COCH_3$, -Cl) is less favourable; the half-wave potentials are higher, the aromatic character is less pronounced as well as is the coplanar structure. These derivatives exhibit a higher neuroleptic activity. The influence of the substituent at the hetero-cycle can also be demonstrated by comparing the electron-donor behaviour of chlorpromazine and 2,3-dichlorpromazine.

The nature and the position of the substituent on the phenothiazine skeleton affects the electron-donor properties of these substances, which can be linked to their pharmacological activity. It was stated that the nitrogen atom is an electron donor in the first oxidation step [9, 10]. This suggestion is in contradiction with the ionization energy of sulphur and nitrogen. If we assume electron influences only, it is possible to determine whether the oxidation occurs at sulphur or nitrogen by correlating the half-wave potentials of the first anodic step with the Hammett constants. We correlated the half-wave potentials for the radical cation formation from simple 2-substituted derivatives and phenothiazine pharmaceuticals (with unbranched propylene side chain between the two nitrogen atoms and substituted in position 2). A good correlation with σ_p values (***Fig. 3***) and a less significant one with σ_m constants for both the groups of derivatives in acid AN medium suggested that sulphur atom, according to the ionization energy, is the electron donor [14, 15] (this conclusion may differ from the equilibrium state characterized by the EPR spectra described in the literature [9, 10]. A good correlation between the half-wave potentials and E_{HOMO} was also achieved.

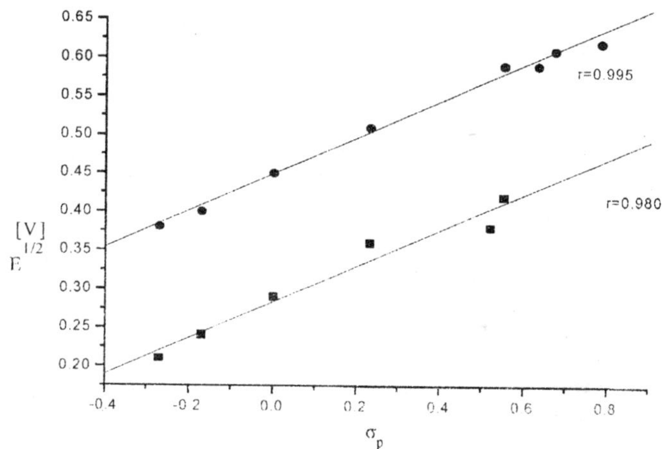

Fig. 3: **Relationship between the half-wave potential for the first oxidation step $E_{1/2}$ [V] and the Hammett constants σ_p**
acid AN medium (0.1M $HClO_4$)
• phenothiazine pharmaceuticals: promazine, chlorpromazine, prochlorperazine, perphenazine, fluphenazine, trifluoperazine, thioproperazine, periciazine and methopromazine$, methylpromazine$, nitropromazine$
▪ phenothiazine and simple derivatives substituted in position 2 with: chlorine, methoxy, trifluoromethyl, acetyl and methyl$ group
$values taken from the literature [9, 10]

Diffusion of the parent molecule to the electrode surface is necessary for the phenothiazine radical cation formation. The diffusion depends on the reaction medium and temperature; the substituent exerts no influence because of the size of the phenothiazine molecule. Once the molecule arrives at the electrode surface, one electron is exchanged. The thermodynamics of the electron-transfer process depends on the structure and the position of the substituent. The kinetics of this process are not influenced by the substituent and depends on the reaction medium only, provided that the acid concentration is low (0.02-0.5 mol dm^{-3}). The substituent influences the kinetics when the acid concentration is high (0.9-1.5 mol dm^{-3}) [18].

Oxidation of the radical cation

Phenothiazine and its derivatives substituted at the phenothiazine nucleus were oxidized in two steps [14]. The radical cation, produced in the first anodic step, was oxidized to the dication. Both the electrode steps were reversible and the charge-transfer rate increased with increasing acid concentration. Reversibility was confirmed by CV; the same results were obtained as in the first electrode step. The peak potentials of the second anodic step shifted to the more positive values with increasing acid concentration; the electroxidation of the radical cation was less favourable. The differences in the peak potentials between the two oxidation steps also increased and the equilibrium constants for the disproportionation of the radical cation were calculated. They were of the order of 10^{-7}-10^{-9} [17]. The dication was more electrophilic than the radical cation and a homogeneous follow-up reaction was probable. Substituents in position 2 made the sulphur atom active and position 3 in thiazine system the most reactive; 3-hydroxy-derivatives were the main final oxidation products. They are known to interact with melanine and cause skin photosensitivity.

In contrast to the substances of the above mentioned group, substitution at the heterocyclic nitrogen atom led to three or four electroxidation processes (***Fig. 2***); it depended on a combined effect of water and acid [13,15,16]. The radical cation was far more reactive and was irreversibly oxidized to the dication. The dication was unstable and was attacked by the acid or water with formation of a protonated sulphoxide species. The higher electrooxidation potentials apparently corresponded to the sulphoxide oxidation. The peak potentials depended on the acid concentration.

A good correlation with the half-wave potentials of 2-substituted derivatives and phenothiazine pharmaceuticals for the second oxidation step with Hammett constants σ_p was achieved [17].

Several exceptions were observed. The mechanism of the second oxidation step of thioridazine and levomepromazine was quasireversible (v = 10-17 V s^{-1}, acid concentration 0.5-1.3M $HClO_4$). These pharmaceuticals have electron donating substituents in position 2 ($-OCH_3$, $-SCH_3$); the charge delocalisation in conjugated systems is higher. The formation of the radical cation and the dication was thus easier (the half-wave and peak potentials were shifted to significantly lower values) and the reaction intermediates had a higher stability [17].

The radical cation of chlorpromazine was prepared in an independent chemical way. It was oxidized in two irreversible steps corresponding to the second and third electrode step. The reduction of the radical cation corresponded to the curves for the first electrode step. The radical cation formation was verified and its electrochemical oxidation products were followed [17].

9.2.2.3 CONCLUSIONS

Our investigation was carried out to collect more data on the electron donor properties and heterogeneous kinetics of phenothiazines. If an electrode surface can be regarded as a very simple model of a biological receptor, the details of the electrode reaction mechanism of phenothiazines are interesting. They can clarify the nature of the phenothiazine-receptor/enzyme interactions and thus also the environmental effects on the metabolism of these drugs. The usefulness of electrochemical methods depends on the correlations among the metabolic, enzymatic and electrochemical oxidation pathways. The conclusions are indirect, but the results obtained in this *in vitro* approach can provide a useful insight into the mechanism of their biological action.

9.2.3 Thiobenzamides

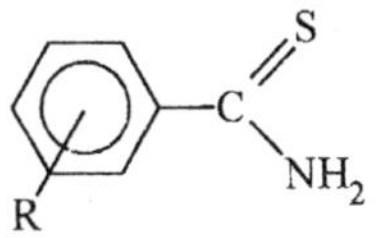

Substituted thiobenzamides exhibit activity against *Mycobacterium tuberculosis* and have been studied in many laboratories as potential antituberculotic drugs. However, thiobenzamides cause undesirable side effects, mainly hepatotoxicity, attributed to a metabolic S-oxidation. It is known that their antituberculotic activity is associated with their electronic properties, particularly with electron density at the sulphur atom.

Our investigation was aimed at a study of the electrochemical oxidation of thiobenzamides to obtain a model for the enzymatic oxidation of these substances. The effect of the substituent position and structure was also studied. The electrochemical oxidation was carried out in an AN medium because of the poor solubility of these substances in aqueous media.

The electrochemical oxidation of thiobenzamides in an anhydrous AN medium containing 0.1M sodium perchlorate as the supporting electrolyte was a totally irreversible and complex process [20]. The anodic peaks were asymmetrical and the CV criteria demonstrated irreversibility. The number of electrons exchanged, found by potentiostatic coulometry, was higher than one. An acid AN medium was thus used.

The best results were obtained in 0.1-0.01M $HClO_4$ in AN. The anodic voltammograms were similar for all the derivatives and the differences between the half-wave potentials and peak potentials depended on the position and structure of the substituent. The Levich and current functions were constant in a range from 2×10^{-5} to 8×10^{-4}M, as were the half-wave and peak potentials. Irreversibility was demonstrated by cyclic voltammetry. Cyclic voltammograms exhibited a single irreversible anodic peak (no cathodic peak was observed), as illustrated in ***Fig. 4***.

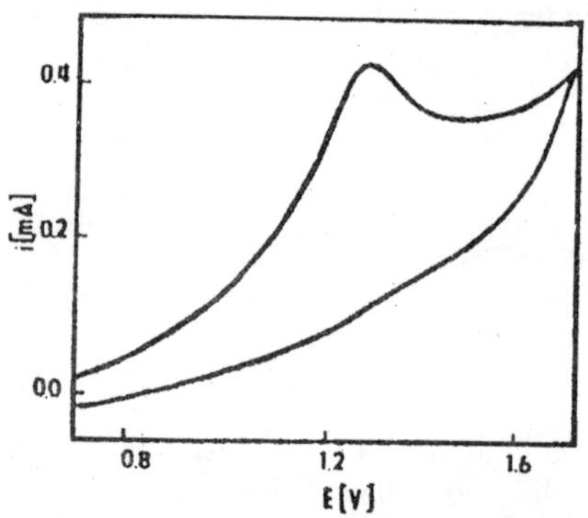

Fig. 4: **Cyclic voltammogram of the anodic oxidation of 4-chlorthiobenzamide ($c = 1 \times 10^{-4}$ mol dm^{-3}) in an AN solution of 0.1M $NaClO_4$**

Concentration of $HClO_4$ 0.1 mol dm^{-3}, temperature 20 °C, $A = 0.288$ cm^2, $v = 0.9$ V s^{-1}

One electron was exchanged per thiobenzamide molecule, as determined by potentiostatic coulometry, within a range of 0.05-0.9M perchloric acid. Constant-potential electrolysis was used to determine the reaction intermediates and products. Hydrogen sulphide was found in the gas released and the EPR spectrum indicated the formation of a radical during the oxidation. From the above results, it was clear that the reaction intermediate was a radical and was converted into an electrochemically inactive product, 2,5-diphenyl-thiodiazole (or a corresponding substituted derivative):

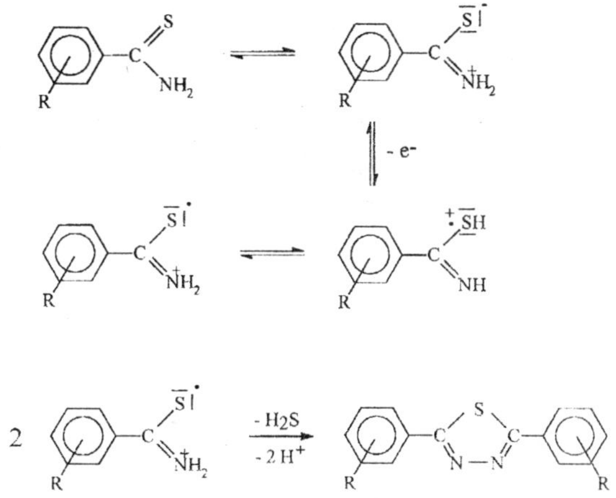

- H_2S
- $2H^+$

This substance was isolated and identified using constant-potential electrolysis of thiobenzamide solution, followed by TL chromatography, UV and mass spectrometry. Nitrile and other products were also isolated.

The half-wave potentials could be correlated with the appropriate Hammett and Brown constants. 4-Methoxy thiobenzamide was most easily oxidized and also exhibited the highest antituberculotic activity.

The charge-transfer coefficients for the radical cation formation were determined in the presence of $HClO_4$ in AN medium, obtaining an α value of approx. 0.2 in the presence of 0.02M $HClO_4$ in AN which increased with increasing acid concentration up to 0.4 (in 0.9M $HClO_4$ in AN) [17].

In addition to the simple thiobenzamide derivatives, *N*-phenylthiobenzamides (thiobenzanilides) were studied [21, 22].

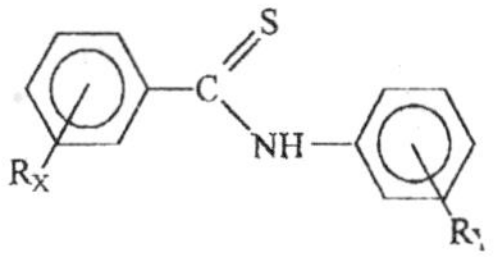

These substances were substituted both in the thioacyl and anilide part of the molecule. It was found that an introduction of the pnenyl group into the molecule caused a decrease in the half-wave and peak potential values (by ca. 0.2V); the electron donor properties were enhanced. Similarly as in the case of simple thiobenzamides, the methoxy group caused a decrease in the half-wave potential values, while the opposite occurred with the chlorine or bromine substitution. In the case of thiobenzanilides, the changes in the half-wave potentials due to substitution were smaller than those for simple thiobenzamides.

The half-wave potentials of simple derivatives were correlated with the Hammett constants using two-parameter regression [23]:

$$E_{1/2} = 0.114\,\sigma_1 + 0.214\,\sigma_2 + 1.038$$

where σ_1 is the constant for the electron effects of the substituents at the thioacyl part of the molecule and σ_2 for the substituents at the anilide part of the molecule. Substituents with electron-acceptor properties caused an increase in the half-wave potentials of the electron transfer, whereas substituents with electron donor properties decreased them. It could be concluded that substitution at the anilide part of the molecule was more important than that at the thioacyl part.

The oxidation was again totally irreversible. The anodic wave and peaks were asymmetric. However, the peak currents and the limiting diffusion currents were proportional to the derivative concentrations over a range of 1×10^{-5}-8×10^{-4} mol dm^{-3}. The number of exchanged electrons in anhydrous AN medium, determined by potentiostatic coulometry, was 1.6-1.8. However, in an acidic AN medium a decrease was observed and one electron was exchanged ($\pm$ 2 %) in a range of acid concentration of 0.05-0.3 mol dm^{-3}.

We tried to isolate and identify the products and intermediates of the electrode reaction and to solve the mechanism of the electrode reaction [22]. Constant-potential electrolysis was used. The released gas contained H_2S; its quantity was determined by precipitation with CdS. It was found that the amount of H_2S corresponded to ca. 40 % of the parent thiobenzanilide. A HPLC analysis of methanol and chloroform extracts of the evaporation residue of the reaction mixture indicated two main products. They were identified by mass spectrometry as benzanilide and 2-phenylbenzthiazole. The identification was further verified by determining the melting points and by employing UV, IR spectra and chromatographic characteristics. The total isolated amount was 55 % of 2-phenylbenzthiazole and 37 % of benzanilide related the amount of thiobenzanilide.

Our results indicated that the electrochemical oxidation of thiobenzanilide is a complex process in which an one electron transfer is complicated by at least two homogeneous reactions.

The hepatotoxicity of thiobenzamides can be caused by H_2S, formed during their metabolism in the liver.

9.3 Conclusions

The usefulness of electrochemical modelling of biological reactions is apparent. This chapter is largely concerned with our work on the mechanism of electrochemical oxidation of biologically and environmentally important substances and is a good example of the way in which electrochemistry can provide unique insights into the mechanisms of biological - enzymatic or metabolic - reactions.

It is quite clear from the data presented in this chapter that the electrochemical oxidation of the studied substances is complex. The electrochemical results yield a great deal of information concerning the standard potentials, reaction intermediates and products.

9.4 References

1. Eckert G. M., Gutmann F., Keyzer K.: *Electropharmacology*. CRC Inc., Boca Raton (1990).
2. Zuman P., Wawzonek S.: *Nonaqueous Solvents in Organic Electrochemistry*. In: *The Chemistry of Nonaqueous Solvents, vol. VA* (Lagowski J. J., Ed.). Academic Press, New York 1978.
3. Pacáková V., Němec I.: J. Chromatogr. *148*, 273 (1978).
4. Weber J. B.: Residue Rev. *32,* 93 (1970).
5. Khan S. U.: *Pesticides in the Soil Environmental*. Elsevier, Amsterdam, 1980.
6. Forrest I. S., Carr C. J., Usdin E., Eds: *Phenothiazines and Structurally Related Drugs*. Raven, New York (1974)
7. Tozer T. N., Tuck L. D., Craig J.C.: J. Med. Chem. *12*, 294 (1969).
8. Waisser K., Odlerová Ž., Valchář M., Němec I., Zimová N.: Folia Pharmaceutica Universitatis Carolinae *XVIII,* 41 (1995).
9. Billon J. P.: Bull. Soc. Chim. France 1923 (1961).
10. Billon J. P.: Ann. Chim. (Paris) *7*, 183 (1962).
11. Cauquis G., Deronzier A., Serve D., Vieil E.: J. Electroanal. Chem. *60*, 205 (1975).
12. Cauquis G., Deronzier A., Lepage J. L., Serve D.: Bull. Soc. Chim. France *295*, 303 (1977).
13. Němec I., Šulcová N., Waisser K.: Českoslov. Farm. *28*, 59 (1979).
14. Šulcová N., Němec I., Waisser K., Kies H. L.: Microchem. J. *25*, 551 (1980).
15. Zimová-Šulcová N., Němec I., Waisser K., Kies H. L.: Microchem. J. *32*, 33 (1985).
16. Zimová N., Němec I., Ehlová M., Waisser K.: Collect. Czech. Chem. Commun. *55*, 63 (1990).

17. Zimová N.: PhD Thesis, Faculty of Natural Science, UK, Prague 1985.
18. Zimová N., Němec I., Waisser K.: *Electrochemical Study of Phenothiazines and its Derivatives in Acid Acetonitrile Medium*. In: *Biological and Chemical Aspects of Thiazines and Analogs* (Barbe J., Keyzer H., Soyfer J. C., Eds.). Enlight Associates, San Gabriel, California (1995)
19. Nicholson R. S., Shain I.: Anal. Chem. *36*, 706 (1964).
20. Němec I., Artner P., Waisser K., Čeladník M., Palát K.: Sborník 7. sjezdu ČSFS, Hradec Králové, UK Praha 1979.
21. Zimová N., Němec I., Waisser K., Zima J.: Českoslov. Farm. *37*, 198 (1988).
22. Gabriel J., Němec I., Zimová N., Hanuš V., Waisser K.: Electroanalysis *6*, 75 (1994).
23. Zimová N., Němec I., Waisser K.: prepared for publication

10

CHEMICAL AND ELECTROCHEMICAL TRANSFORMATIONS OF 1,2,4-TRIAZINE HERBICIDES

F. Riedl, J. Ludvík, J. Volke, P. Zuman, F. Liška

Abstract

Synthesis and the chemical and electrochemical properties of 1,2,4-triazine herbicides have been investigated and discussed. Electrochemical oxidation and reduction has been used to study the degradation of these compounds. Possibilities of their determination are outlined.

Key Words

1,2,4-triazine, herbicide, synthesis, properties, degradation, electrochemical properties

10.1 Introduction

The plant protection chemicals (herbicides) used at present substantially differ in their structures [1]. Some of them can be derived from phosphoric or phosphonic acid, such as Phosphamidone, from carbonic acid, e. g. Linurone; further compounds are based on aromatics, e. g. Prochloraz, on quinoline, e. g. Oxine, on 4,4'-bipyridine, e.g. Paraquat, and, quite recently on 1,3,5-triazines, e.g. Prometryne, and on asymmetric 1,2,4-triazines.

Our investigation concentrates on the description of chemical and, in particular, on electrochemical properties of 1,2,4-triazine-based herbicides - Metamitrone (I) and Metribuzine (II), which are the active components of preparations Goltix WP 70 and Sencor WP 70 used in agricultural practice.

The papers concerning the above two herbicides are mainly devoted to their synthesis investigation of their degradation under different conditions and to determination of their spectral and, in particular, redox properties. We have been mainly interested in the proof of their presence in various media and in the determination of I and II and their degradation products, as a function of time.

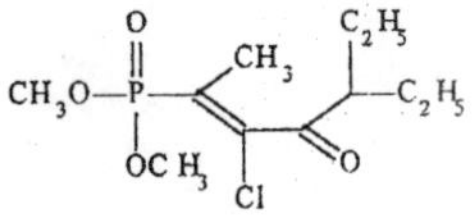

PHOSPHAMIDONE 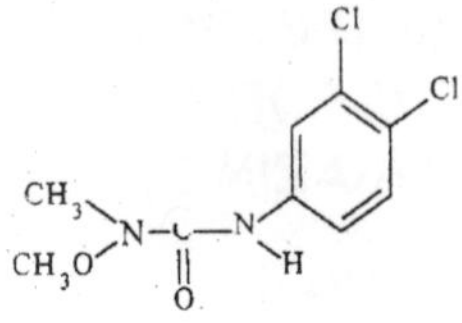 LINURONE 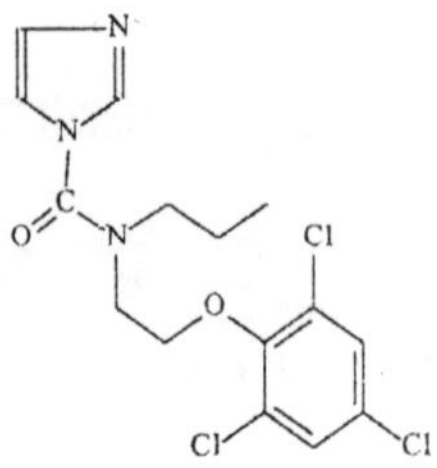 PROCHLORAZ

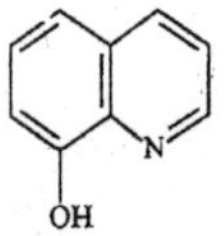

OXINE 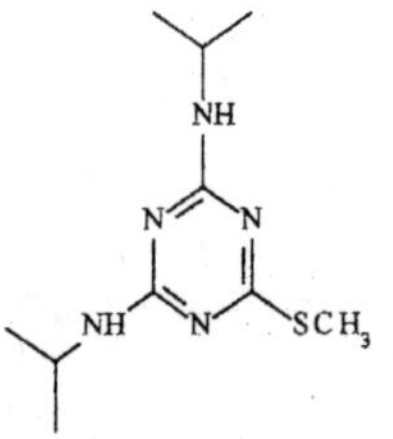 PROMETRYNE 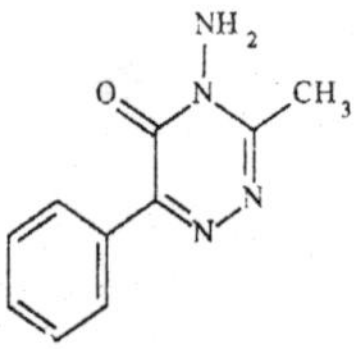 METAMITRONE (I) 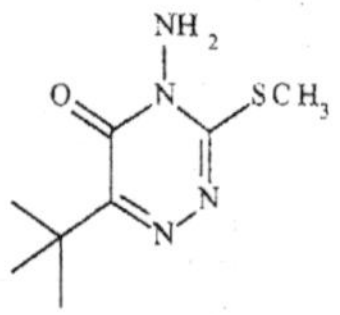 METRIBUZINE (II)

10.2 Synthesis of 1,2,4-triazine-based herbicides

When following the fate of herbicides in the environment, we are interested in the products of their degradation. For this reason we are going to describe here the synthesis of I and II since the degradation products may involve the substances from which they have been synthesized could be found. The synthesis of 1,2,4-triazine herbicides I and II starts from 2-oxoacids and their functional derivatives. In the case of I, these are chlorides, nitriles and esters of phenylglyoxylic acid [2,3] which are first transformed into acetyl-hydrazones and finally to the corresponding hydrazone-hydrazides [4-6] yielding Metamitrone (I). A different procedure is based on a reaction of ethylphenylglyoxylate with acethydrazidine [6].

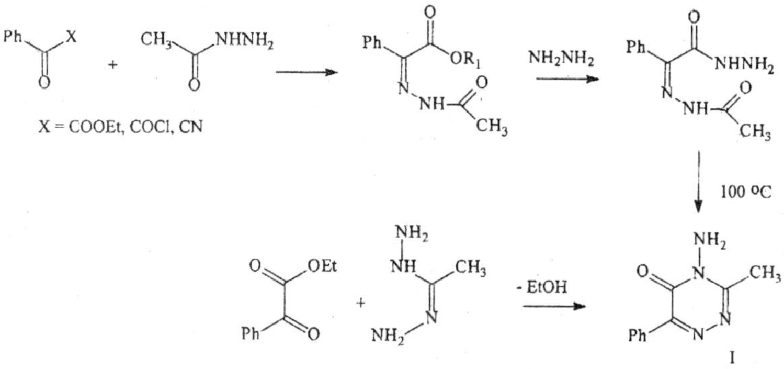

Ph-C(=O)-X + CH3-C(=O)-NHNH2 → Ph-C(=N-NH-C(=O)-CH3)-C(=O)-OR1 --NH2NH2→ Ph-C(=N-NH-C(=O)-CH3)-C(=O)-NHNH2

X = COOEt, COCl, CN

100 °C

Ph-CO-COOEt + NH2-NH-C(CH3)=N-NH2 --- EtOH→ I

The synthesis of Metribuzine (II) starts from *t*-butylglyoxylic acid [7, 8] or from its functional derivatives (amide or N-acetylamide) and the triazine structure results from the cyclocondensation with thiourea dihydrazide. An intermediate is the thiol III which is finally methylated by CH_3I or CH_3Br in alkaline media to II. The individual procedures only differ in obtaining the corresponding 2-oxoacids and their derivatives or in the modification of conditions of the cyclocondensation.

10.3 Chemical properties of 1,2,4-triazine herbicides

The presence of the methylsulphanyl group in position 3 in Metribuzine (II) enables reactions inaccessible in Metamitrone (I) containing a methyl group in position 3. This can be exemplified by the nucleopholic vinyl substitution of the methylsulphanyl group taking place in an addition/elimination mechanism by the action of, e.g., alkanethiols [9], alkanols [10] and amines [11]. An analogous interpretation can be applied in the formation of the hydrolytic product, the dione IV.

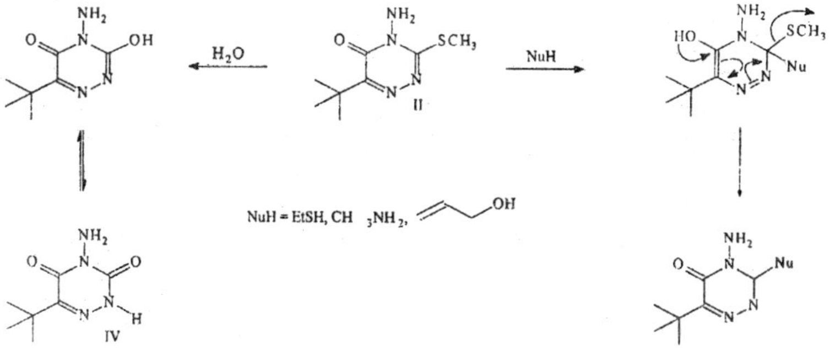

H_2O ← II → NuH

IV

NuH = EtSH, CH_3NH_2, allyl alcohol (OH)

An interesting substitution of the methylsulphanyl group occurs in an acid-catalyzed reaction with dimethylaminotrimethylsilane [12].

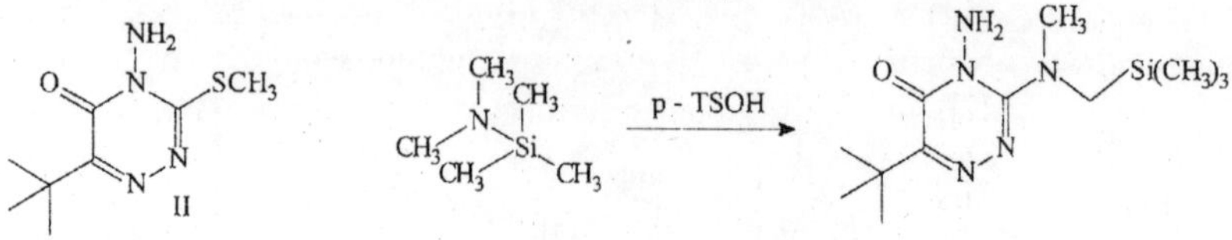

NH2 O N SCH3 N N II CH3 CH3 N CH3 Si CH3 CH3 p - TSOH NH2 CH3 O N N Si(CH3)3 N N

The amino group in position 4 in Metribuzine (II) is sufficiently active in the reaction with acylating agents (sulphenyl chlorides, alkanoyl chlorides) [13-15]; reactions with acetals of glyoxylic acid esters lead to the formation of the corresponding aminals [16, 17], the reaction with ketones to "hydrazones" [18] and, finally, the reaction with formaldehyde yields the corresponding methylol derivative [19]. The reaction with formic acid [20] results in the formylation of the amino group. A reaction with DMSO gives rise to sulphimides [21].

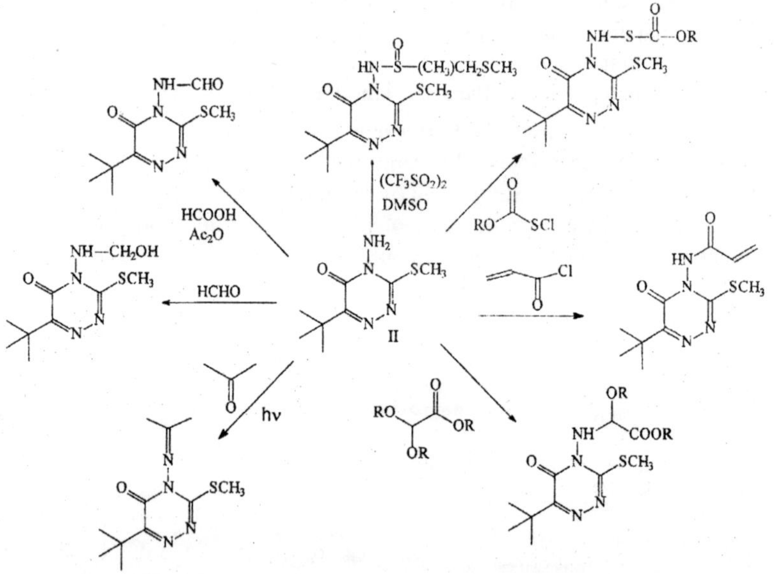

NH—CHO O N SCH3 N N HN—S—(CH3)CH2SCH3 O N SCH3 N N NH—S—C—OR O N SCH3 N N (CF3SO2)2 DMSO HCOOH Ac2O RO SCl O NH—CH2OH O N SCH3 N N HCHO NH2 O N SCH3 N N II Cl O HN O N SCH3 N N hν RO OR OR O OR NH COOR O N SCH3 N N N O N SCH3 N N

Most of the above reactions have been detected in searching for synthetic routes to obtain novel herbicides or, at least, substances exhibiting some herbicidal activity.

10.4 Analytical determination of 1,2,4-triazine herbicides

In the determination of triazine derivatives I and II and for identification of their degradation products, various analytical methods have been applied, such as HPLC [22], GLC [23], GC-MS [24,25] or even calorimetry [26]. The crystal structure of II was determined [27]. The electrochemical procedures based on the results obtained by the present authors are described in detail in Part 10.6.

10.5 Degradation of 1,2,4-triazine herbicides

A relatively great attention has been devoted to the photolysis ($\lambda > 290$ nm) of Metribuzine (II) [28-31]; an analogous investigation of Metamitrone (I) has not been so thorough [32,33]. In both cases, fission of the N-N grouping in position 4 leads to deamination and the main products are triazines V and VI; triazines IV, VII and VIII have been found as by-products. The photolysis was studied both in the crystalline state [33] and in solution (H_2O, CH_3OH, CCl_4 and CH_2Cl_2) [33,34]. The deamination depends [29,35] on the O_2 and H_2O content in the solution and an interaction with HCOO. radicals is assumed in the first reaction step [29].

The determination of quantum yields in the photodegradation is recommended as a criterion of the degradability [36] of the two herbicides. It has been observed that the rate of photolytic degradation of Metamitrone (I) is decreased by increasing the pH-value of the solution over a range from 4 to 8; in photolysis in unbuffered solutions the acidity of the solution increases from pH 7 to 3.7; simultaneously, the UV spectra [32] are changed.

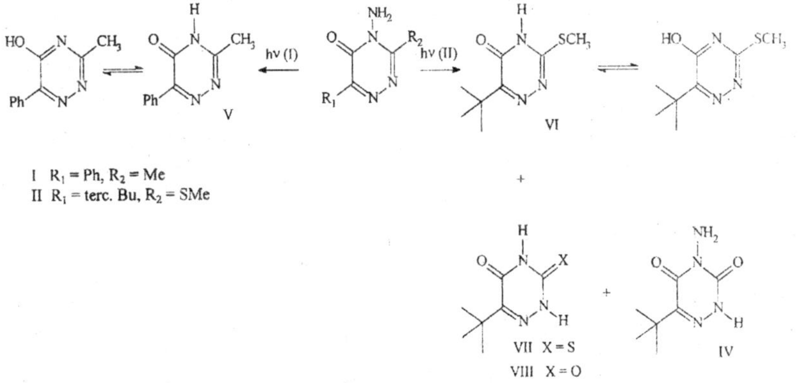

HO N CH$_3$ — Ph N N ⇌ O N(H) CH$_3$ — Ph N N V ← hν (I) — O N(NH$_2$) R$_2$ — R$_1$ N N — hν (II) → O N(H) SCH$_3$ N N VI ⇌ HO N SCH$_3$ N N

I R$_1$ = Ph, R$_2$ = Me
II R$_1$ = terc. Bu, R$_2$ = SMe

+

VII X = S
VIII X = O
+
IV

The course of oxidation of I and II depends on the reactant used, the solvent and on the catalyst [37]. Whereas in the oxidation by Pb(IV)acetate in CH_2Cl_2, the lactam IV results exclusively, the sulphimide IX is formed in DMSO. *t*-Butylhydroperoxide or H_2O_2 react with I or II only in the presence of the catalyst $M(salem)_2$ [where M = Co(II), Mn(II), salem = N,N-disalicylideneethylenediamine] in CH_2Cl_2, and at room temperature they yield quantitatively the deaminated V and VI. In methanolic solutions, the reaction proceeds more slowly and II is transformed into a mixture of IV, VI and VIII. With other catalysts [$MnO(acac)_2$ or $VO(acac)_2$], II is quantitatively transformed into IV. In the oxidation of II, no products has been isolated resulting from an oxidation at the sulphur atom although sulphides are easily oxidized by hydroperoxy compounds.

I R_1 = Ph, R_2 = Me
II R_1 = *terc.* Bu, R_2 = SMe

The oxidation of the above herbicides can serve as a chemical model of their metabolism. In an attempt to interpret the mechanism of oxidative deamination of these two herbicides, the fate of the "eliminated" amino group or the oxidation level of the compound into which the amino group is transformed, remain unknown.

Thus, e.g., the dioxo derivatives VI and VIII resulting from the deamination and hydrolysis of Metribuzine (II) were found in the Hawaian Sea sediments [38] or were identified in photochemical or thermal decomposition on the surface of silica gel [39,40] or of borosilicate glass [40]. The photoinitiated transformations of pesticides performed on silica gel and taking place according to pseudo-first-order kinetic law are a model of their degradation [39].

Biodegradations were followed *in vitro,* in the presence of various strains (e.g., *Cunninghamella echinulata*) and the triazine V was isolated [41] as the main product. In a similar manner, Metribuzine (II) is transformed [42] in presence of peroxysomes obtained from soya beans into triazinone VI. The biodegradation of II *in vivo* was studied on mice in sublethal doses (150-250 mg kg^{-1}); II exhibits hepatotoxicity and the urine of mice contained mercapturic acid derivatives resulting from a reaction between glutathione and Metribuzine-sulphoxide or deaminated Metribuzine-sulphoxide [43].

When applying nitrate-containing fertilizers, one can expect the presence of nitrites in soil. For this reason we studied the reaction of I or II with nitrous acid. In case of I, this reaction has been described and triazinone V was identified as the only product [44]; the further fate of the nitrogen atom, however, was not investigated. In the reaction of II with HNO_2 we also observed an elimination of the amino group and it was possible to prove with the help of IR spectra that the transient intermediates, i. e., azohydroxides, eliminate N_2O and are transformed to the lactams V and VI. Moreover, the formation of CO has been proved: this behaviour can be explained by the decarboxylation of the transiently formed phenylglyoxylic acid or *t*-butyl-glyoxylic acid.

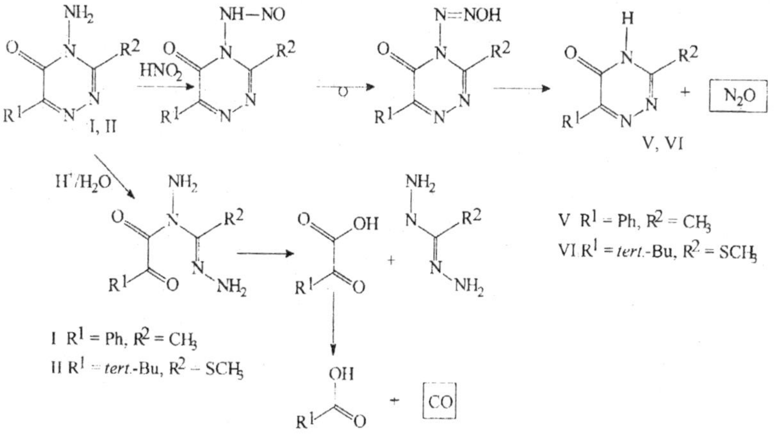

NH₂ | NH–NO | N=NOH | H

HNO_2 | $-O$ | + N_2O

I, II | V, VI

H^+/H_2O | + | + CO

V R^1 = Ph, R^2 = CH_3

VI R^1 = *tert.*-Bu, R^2 = SCH_3

I R^1 = Ph, R^2 = CH_3

II R^1 = *tert.*-Bu, R^2 = SCH_3

10.6 Reducibility and electrochemical investigation of 1,2,4-triazine herbicides

The reduction of I and II has not yet been investigated in detail, although it is well known that the azine derivatives are generally easily reducible. A selective reduction of Metamitrone(I) by sodium tetrahydroborate proceeds in methanolic solutions [35]. The product of such a reaction is the corresponding 3,4-dihydro-1,2,4-triazinone X.

$$\text{I} \xrightarrow[\text{MeOH}]{\text{NaBH}_4} \text{X}$$

A pioneering polarographic investigation of triazine herbicides was published by Polák and Volke [46]. These authors found that the triazines I and II can be determined polarographically (DC and differential-pulse polarography) at concentrations as low as 1×10^{-7} mol l^{-1}, but they did not study the isolation of the reduction products and their identification. The Spanish authors [47] later used differential-pulse polarography in the determination of dissociation constants of I. However, the data obtained from these DPP measurements are insufficient for determining the dissociation constants of Metamitrone (I): such a determination should be preferably based on spectrophotometric measurements in a series of buffers.

Recently, we have studied the electrochemical behaviour of Metamitrone I [48]. As both d.c. polarography (DCP) and cyclic voltammetry (CV) at a hanging mercury drop electrode (HMDE) of compound I in purely aqueous solutions were strongly affected by adsorption phenomena, most informative results were obtained in mixed (30 % v/v acetonitrile) solvents.

At the pH range 0.5 to 8, Metamitrone is reduced in principle in two diffusion controlled two-electron waves (***Fig. 1***, upper curve), which are irreversible and pH-dependent. The number of electrons exchanged in the individual steps was determined either by comparing the heights of the polarographic waves with those of standards with a known electron consumption or by controlled-potential coulometry.

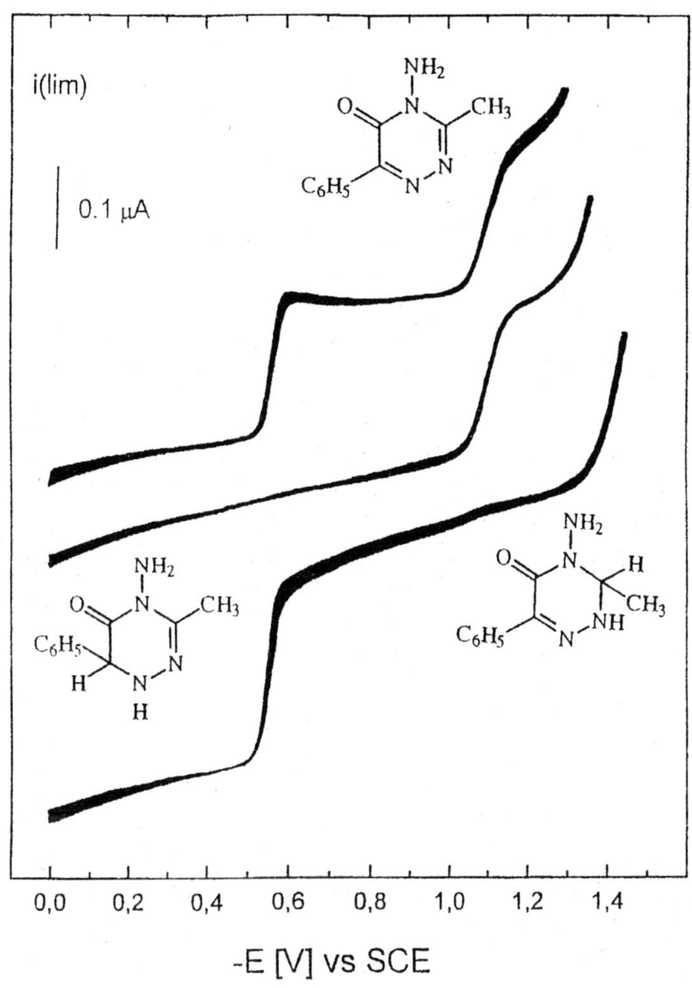

Fig. 1 **DC-polarography of Metamitrone, 1,6-dihydrometamitrone and 2,3-dihydrometamitrone**
pH 4.2, concentration 5×10^{-4} mol l^{-1}

In the case of the first wave the half-wave potential/pH dependence over the whole mentioned pH region is reflected by the straight line. This is in agreement with the $H^+e^-e^-H^+$ mechanism where antecedent protonation occurs. In the case of the second reduction wave the analogous linear pH dependence is observed at pH > 4. At pH < 4 with increasing acidity a new, more negative wave appears indicating an acid catalyzed reaction (tautomeric change or ring opening).

On the other hand, amino substituent at the nitrogen in the position 4 represents a hydrazino (or hydrazide) grouping where an oxidation process can be anticipated. In fact, no anodic wave was observed on Hg electrodes, moreover, even the fast CV did not reveal any cathodic peak neither in the first scan, nor in the subsequent cycles.

In strongly alkaline media (at pH values above 12), an alkaline hydrolysis of triazines I and II proceeds leading to fission of the cycle. This fact is accompanied by the formation of a new anodic wave at about +0.05 V (SCE) and by changes in the UV/VIS spectra where the originally observed band at 310 nm is replaced by a new one at 250 nm through isosbestic points.

When a CPE proceeds at the limiting current of the first wave, the isolated product was identified as 1,6-dihydrometamitron, that means at these potentials the 1,6-duble bond is reduced as the first electroreduction step of Metamitrone. The obtained 1,6-dihydroderivative is then reduced in one more negative step which fully corresponds to the second wave of Metamitrone (***Fig. 1***, middle curve).

The product of sodium borohydride reduction, 2,3-dihydrometamitron, being reduced exhibits only one electroreduction step corresponding to the 1,6-double bond reduction, whereas the more negative wave of Metamitrone can be attributed to the further reduction process initiated by the reduction of 2,3-azomethine bond.

The first reduction wave of Metamitrone exhibits a decrease of the limiting current in the pH region 2.5 to 6.5, which reaches approximately 80 % of its original value. The similar effect exhibits also the 2,3-dihydrometamitron. This dip on the pH dependence of the limiting current is attributed to the covalent hydration of 1-6- C=N double bond in this pH range. The hydrated adduct is not more electroactive thus its presence diminishes the actual concentration of electroactive Metamitrone and causes the lowering of the limiting current. The shape of the decrease indicates that the dehydration can be both acid and base catalyzed: at pH < 2 and pH > 7 the current reaches the theoretical level.

Metamitrone cannot undergo tautomeric changes antecedent the first electron transfer. On the other hand, tautomeric changes are in principle possible in the case of both isolated dihydrometamitrone derivatives.

From the results obtained until today one can conclude that:

The first electroreduction wave represents the reduction of the 1,6-azomethine bond with antecedent protonation of the N(1). The second electroreduction wave corresponds to the reduction of 1,6-dihydro-metamitron where a precedent protonation is involved. For this reduction process the presence of the 2,3-azomethine bond is necessary.

The borohydride attacks only the 2,3-double bond of Metamitrone, thus there is a different mechanism of the electrochemical and chemical reduction of Metamitrone. This fact can be attributed to a difference in reaction conditions: Whereas the electrochemical reduction proceeds as a heterogeneous process involving a protonated molecule of (I), in the chemical reduction of the borohydride reacts with an uncharged molecule in a homogeneous phase.

Reduction under conditions of DC-polarography is complicated by acid-base hydration-dehydration equilibria and another (tautomeric or ring-opening) processes.

The molecule of Metamitrone with two conjugated azomethine groups represents a system with two redox active centres. In this view, there is an unusually small interaction between them because the activity of the C=N double bond in the position 1,6 is not affected by the presence or absence of the C=N bond in the position 2,3 and vice versa. This problem is under further investigation.

10.7 References

1. A List of Permitted Preparations for Plant Protection, Ministry of Industry, the Czech Republic, 1993.
2. Findiesen K., Draber W., Schwarz H.: Ger. Offen. 2,528,211; 20. Jan 1977; [CA 87:5666c].
3. Schwarz H., Kraetzer H., Singer R. J.: Ger. Offen. 2,828,011; 10. Jan 1980; [CA 92:215079t].
4. Draber W., Timmler H., Dickore K., Donner W.: Justus Liebigs Ann. Chem. (12), 2206 (1976).
5. Timmler H., Draber W.: Ger. 2,366,215; 19. Jul 1979; [CA 91:157472g].
6. Neunhoeffer H., Degen H. J.: Ger. Offen. 2,556,835; 30. Jun 1977; [CA 87:117910j].
7. Jackman D. E.: J. Heterocycl. Chem. *27*, 1053 (1990).
8. Staicu S. A., Caproiu M. T., Mihailescu M. N., Constantin D. M., Zenide D. A.: Rom. 66,584; 15. Feb 1979; [CA 95:115615h].

9. Jackman D. E., Westphal D. B., Shidt T.: Eur. Pat. Appl. EP 168,766; 22. Jan 1986; [CA 104:224920d].
10. Roy W., Santel H. J., Schmidt R. R.: Ger. Offen. DE 3,510,792; 25. Sep 1986; [CA 106:32128h].
11. Tocker S.: U. S. US 4,632,694; 30. Dec 1986; [CA 106:133813w].
12. Murler K. H., Dickore K., Eue L., Saqntel H. J., Schidt R R.: Ger. Offen. DE 3,510,789; 25. Sep 1986; [CA 106:50261p].
13. Pilgram K. H., Bozarth G. A.: U. S. US 4,589,913; 20. May 1986; [CA 105:56358b].
14. Shell Internationale Reserch Maatschappij B. V.: Jpn. Kokai Tokkyo Koho JP 61,221,182 [86,221,182]; 1. Oct 1986; [CA 106:84656m].
15. Mc Cormic C. L., Zhang Z. B., Anderson K. W.: Polym. Prepr. *24*, 364 (1983).
16. Draber W., Dickore K., Timmler H., Eue L., Schidt R. R.: Ger. Offen. 2,620,370; 17. Nov 1977; [CA 88:62422q].
17. Draber W., Dickore K., Timmler H., Eue L., Schidt R. R.: Ger. Offen. 2,540,958; 17. Mar 1977; [CA 87:135416g].
18. Mueller K. H., Roy W., Santel H. J., Schmidt R. R.: Ger. Offen. DE 3,510,786; 25. Sep 1986; [CA 106:33127g].
19. Rocker S.: U. S. US 4,614,799; 30. Sep 1986; [CA 106:28839m].
20. Roy W., Santel H. J., Schmidt R. R.: Ger. Offen. DE 3,510,785; 25. Sep 1986 [CA 106:33129j].
21. Mueller K. H., Roy W., Eue L., Santel H. J.: Ger. Offen. DE 3,510,794; 25. Sep 1986; [CA 106:33130c].
22. Pepperman A. B., Kuan J. W.: J. Liq. Chromatogr. *15*, 819 (1992).
23. Saxton W. L.: J. Chromatogr. *393*, 175 (1987).
24. Matuschek G., Ohrbach K. H., Kettrup A.: Thermochim. Acta *190,* 111 (1991).
25. Kienhuis P. G. M.: J. Chromatogr. *647*, 39 (1993).
26. Donnely J. R., Drewes L. A., Johnson R. L., Munslow W. B., Knap K. K., Sovocool G. W.: Thermochim. Acta *167*, 155 (1990).
27. Aliev Z. G., Atovmyan L. O., Kartsev V. G.: Zh. Strukt. Khim. *30,* 162 (1989).
28. Herrmann M., Parlar H., Korte F.: Chem.-Ztg. 106 373 (1982).
29. Herrmann M., Parlar H., Korte F.: Z. Naturforsch., B: Anorg. Chem., Org. Chem. *37 B*, 117 (1982).
30. Parlar H., Pletsch B.: Chemosphere *17*, 2043 (1988).
31. Draber W., Dickore K., Timmler H.: Dokl. Soobshch.-Mezhdunar. Kongr. Zashch. Rast., 8th 3, 203 (1975).
32. Schidt R. R.: Meded. Fac. Landbouwwet. Rijksuniv. Gent *41*, 1289 (1976).
33. Pape B. E., Zabik M. J.: J. Agr. Food Chem. *20*, 72 (1972).
34. Bartl P., Korte F.: Chemosphere *4*, 169 (1975).
35. Bartl P., Parlar H., Korte F.: Z. Naturforsch., B: Anorg. Chem., Org. Chem. *31 B*, 1122 (1976).
36. Parlar H., Korte F.: Chemosphere *8*, 797 (1979).
37. Nakayama Y., Sanemitsu Y., Yoshioka H., Nishinaga A.: Tetrahedron Lett. *23*, 2499 (1982).
38. Miles C. J., Leong G., Dollar S.: Bull. Environ. Contam. Toxicol. *49*, 179 (1992).
39. Hulpke H., Stegh R., Wilmes R.: Pestic. Chem.: Hum. Welfare Environ., Proc. Int. Congr. Pestic. Chem., 5th 3, 323 (1983).

40. Bartl P., Korte F.: Chemosphere *4*, 173 (1975).
41. Engelhardt G., Wallnoefer P. R.: Chemosphere *7*, 463 (1978).
42. Fedtke C.: Naturwiss. *70*, 199 (1983).
43. Bleek M. S., Smith M. T., Casida J. E.: Pestic. Biochem. Physiol. *23*, 123 (1985).
44. Draber W. et al.: Justus Liebigs Ann. Chem. *2206*, 2218 (1976).
45. Draber W., Timmler H., Eue L., Schmidt R. R.: Ger. Offen. 2,346,936; 3. Apr 1975; [CA 83:97385d].
46. Polák J., Volke J.: Českosl. Farm. *32*, 282 (1983).
47. Goicolea A., Arranz J. F., Barrio R. J.: Anales de Quimica *87*, 163 (1991).
48. Ried P., Ludvík J., Liška P., Zuman P.: J. Heterocycl. Chem. 33, 1, (1996).

LIST OF CONTRIBUTORS

Jiri Barek

UNESCO Laboratory, Department of Analytical Chemistry, Faculty of Natural Sciences, Charles University, Albertov 2030, 128 40 Prague 2, Czech Republic

Milan Fedurco

UNESCO Laboratory, J. Heyrovsky Institute of Physical Chemistry, the Czech Academy of Sciences, Dolejskova 3, 182 23 Prague 8, Czech Republic

Michael Heyrovský

UNESCO Laboratory, J. Heyrovsky Institute of Physical Chemistry, the Czech Academy of Sciences, Dolejskova 3, 182 23 Prague 8, Czech Republic

Robert Kalvoda

UNESCO Laboratory, J. Heyrovsky Institute of Physical Chemistry, the Czech Academy of Sciences, Dolejskova 3, 182-23 Prague 8, Czech Republic

Miloslav Kopanica

UNESCO Laboratory, J. Heyrovsky Institute of Physical Chemistry, the Czech Academy of Sciences, Dolejskova 3, 182 23 Prague 8, Czech Republic

Frantisek Liska

Institute of Organic Chemistry, Institute of Chemical Technology, Technická 5, 166 28 Prague 6, Czech Republic

Jiri Ludvik

UNESCO Laboratory, J. Heyrovsky Institute of Physical Chemistry, the Czech Academy of Sciences, Dolejskova 3, 182 23 Prague 8, Czech Republic

Ivan Nemec

UNESCO Laboratory, Department of Analytical Chemistry, Faculty of Natural Sciences, Charles University, Albertov 2030, 128 40 Prague 2, Czech Republic

Ladislav Novotný

UNESCO Laboratory, J. Heyrovsky Institute of Physical Chemistry, the Czech Academy of Sciences, Dolejskova 3, 182 23 Prague 8, Czech Republic

Frantisek Opekar

UNESCO Laboratory, Department of Analytical Chemistry, Faculty of Natural Sciences, Charles University, Albertov 2030, 128 40 Prague 2, Czech Republic

Vera Pacáková

UNESCO Laboratory, Department of Analytical Chemistry, Faculty of Natural Sciences, Charles University, Albertov 2030, 128 40 Prague 2, Czech Republic

Frantisek Riedl

UNESCO Laboratory, J. Heyrovsky Institute of Physical Chemistry, the Czech Academy of Sciences, Dolejskova 3, 182 23 Prague 8, Czech Republic

Zdenek Samec

UNESCO Laboratory, J. Heyrovsky Institute of Physical Chemistry, the Czech Academy of Sciences, Dolejskova 3, 182 23 Prague 8, Czech Republic

Lubomir Serák

UNESCO Laboratory, J. Heyrovsky Institute of Physical Chemistry, the Czech Academy of Sciences, Dolejskova 3, 182 23 Prague 8, Czech Republic

Ivana Sestáková

UNESCO Laboratory, J. Heyrovsky Institute of Physical Chemistry, the Czech Academy of Sciences, Dolejskova 3, 182 23 Prague 8, Czech Republic

Karel Stulik

UNESCO Laboratory, Department of Analytical Chemistry, Faculty of Natural Sciences, Charles University, Albertov 2030, 128 40 Prague 2, Czech Republic

Stanislav Vavricka

Department of Physical and Macromolecular Chemistry, Faculty of Natural Sciences, Charles University, Albertov 230, 128 40 Prague 2, Czech Republic

Jiri Volke

UNESCO Laboratory, J. Heyrovsky Institute of Physical Chemistry, the Czech Academy of Sciences, Dolejskova 3, 182 23 Prague 8, Czech Republic

Jiri Zima

UNESCO Laboratory, Department of Analytical Chemistry, Faculty of Natural Sciences, Charles University, Albertov 2030, 128 40 Prague 2, Czech Republic

Nad'a Zimová

UNESCO Laboratory, Department of Analytical Chemistry, Faculty of Natural Sciences, Charles University, Albertov 2030, 128 40 Prague 2, Czech Republic

Petr Zuman

Department of Chemistry, Clarkson University, Potsdam 13699-5810, N.Y., USA